HOW TO REBUILD

VW AIR-COOLED ENGINES

1961–2003

Prescott Phillips

S-A DESIGN

CarTech®

CarTech®

CarTech®, Inc.
6118 Main Street
North Branch, MN 55056
Phone: 651-277-1200 or 800-551-4754
Fax: 651-277-1203
www.cartechbooks.com

Edit by Bob Wilson
Layout by Connie DeFlorin
Cover Image Provided by Roy Henning

ISBN 978-1-61325-433-2
Item No. SA221

Library of Congress Cataloging-in-Publication Data

Names: Phillips, Prescott, 1962- author.
Title: VW air-cooled engines : how to rebuild / Prescott Phillips.
Description: Forest Lake, MN : CarTech, Inc., [2019]
Identifiers: LCCN 2019003189 | ISBN 9781613254332
Subjects: LCSH: Volkswagen automobiles–Motors–Maintenance and
  repair–Handbooks, manuals, etc. | Volkswagen
  automobiles–Motors–Modification–Handbooks, manuals, etc. |
Air-cooled engines–Maintenance and repair. | LCGFT: Handbooks and
manuals.
Classification: LCC TL215.V6 P48 2019 | DDC 629.25/040288–dc23
LC record available at https://lccn.loc.gov/2019003189

Written, edited, and designed in the U.S.A.
Printed in China

10 9 8 7 6 5 4 3 2

DISTRIBUTION BY:

Europe
PGUK
63 Hatton Garden
London EC1N 8LE, England
Phone: 020 7061 1980 • Fax: 020 7242 3725
www.pguk.co.uk

Australia
Renniks Publications Ltd.
3/37-39 Green Street
Banksmeadow, NSW 2109, Australia
Phone: 2 9695 7055 • Fax: 2 9695 7355
www.renniks.com

Canada
Login Canada
300 Saulteaux Crescent
Winnipeg, MB, R3J 3T2 Canada
Phone: 800 665 1148 • Fax: 800 665 0103
www.lb.ca

CONTENTS

About the Author

Prescott Phillips has been building stock and high-performance Volkswagen engines and transmissions since 1981. A semiretired Journeyman Tool and Die Maker, he always strives for perfection and has the reputation for building strong, reliable powerplants. His love for drag racing and Volkswagens led him to build some of the fastest Beetles in the Midwest. Prescott's team, Underdog Racing, was formed in 1994 and campaigned an all-motor 11-second 1956 Oval Window Bug for many years. In 2008, Prescott built a 10-second 1960 VW Pro Stock Beetle to campaign in the ECPRA series and has done rather well. Recently, he built a 1953 street-legal Drag Bug to race in the 2017 *Hot Rod* Drag Week and finished the grueling 1,000-mile, 5-racetrack week with a best ET of 12.45 at 106 mph, averaging more than 20 mpg!

Acknowledgments

I always thought that it would be a shame for all my knowledge of VW engines to disappear once my time on the planet came to an end. I never thought that way when I was a younger man, but when people start asking you when you plan on retiring, these thoughts become more frequent. Then, when the opportunity presented itself to actually put my decades of VW experience to print, I became apprehensive. But as the famous saying goes, "How hard can it be?" I was willing to give it my best shot.

This book was earmarked to be written by a number of other experts but never came to be. Dedication of time is a huge factor in these things. I found out through the local VW grapevine that this book had run into a roadblock. I sent an email to CarTech Books stating I'd be interested in writing a book on rebuilding a stock air-cooled VW engine. That same day, Bob Wilson of CarTech Books contacted me, and I had to convince him I was serious about writing the book and had the knowledge to put forth a good product. Kudos to Bob for his faith in me.

I'm not a photographer, and these publications need professional-quality photographs, so I went on a quest to find someone who was willing to work with me. The pay was going to be dismal and the hours long. I reached out to a couple close friends and hit the jackpot. Marybeth Kiczenski (shelbydiamondstar.com) was recommended to me. We hit it off from the very first photo session. It was a learning experience for both of us. Her professionalism and easygoing personality were the perfect combination for a successful relationship. Her enthusiasm for the project really shows. Without her help, there's no doubt in my mind that this project would never have come to be.

My oldest brother, Peter, was instrumental in my early interest in Volkswagens. In my early teens, he was constantly working on his 1968 Beetle and left hot VW magazines lying around the house for me to pick up and fantasize about building my own Bug. The engine in this book was pulled out of his 1970 Baja Bug, and the fresh rebuild will be installed back in it after the bodywork and paint are finished. He painted my first couple Volkswagens, and I owe him so much more than just a fresh powerplant for his.

All of my friends had a small hand in the writing of this book but none more than one of my oldest friends, Peter Karempelis, or Peter K as he is known. He shared his tools and parts for many of the photos in this book. We were mid photo shoot when I realized we didn't have some very important items to capture. A quick call to Peter K and he hand delivered them to save the day. Friends like that are invaluable and make projects like this much more rewarding.

I contacted a few VW aftermarket manufacturers and Volkswagen of America. None of them were willing to help with the development of this publication except for one. I contacted EMPI Inc., and Erica Cooper quickly responded with a "How can I help?" attitude. She provided me with all the information I asked for, and it was refreshing to know that someone in the VW aftermarket was willing to help.

Finally, I have to thank my beautiful wife, Paula. Not only did she have to put up with my usual shenanigans but the extra project of writing this book really taxed our together time. I told her I would make it up to her by taking her on vacation next year. I just need to find out what the dates are for the VW Pro Stock drag race in South Carolina.

WHAT IS A WORKBENCH® BOOK?

This Workbench® Series book is the only book of its kind on the market. No other book offers the same combination of detailed hands-on information and close-up photographs to illustrate rebuilding and modifying. Rest assured, you have purchased an indispensable companion that will expertly guide you, one step at a time, through each important stage of the rebuilding process. This book is packed with real-world techniques and practical tips for expertly performing rebuild procedures, not vague instructions or unnecessary processes. At-home mechanics or enthusiast builders strive for professional results, and the instruction in our Workbench® Series books help you realize pro-caliber results. Hundreds of photos guide you through the entire process from start to finish, with informative captions containing comprehensive instructions for every step of the process.

The step-by-step photo procedures also contain many additional photos that show how to install high-performance components, modify stock components for special applications, or even call attention to assembly steps that are critical to proper operation or safety. These are labeled with unique icons. These symbols represent an idea, and photos marked with the icons contain important, specialized information.

Here are some of the icons found in Workbench® books:

Important!
Calls special attention to a step or procedure, so that the procedure is correctly performed. This prevents damage to a vehicle, system, or component.

Save Money
Illustrates a method or alternate method of performing a rebuild step that will save money but still give acceptable results.

Torque Fasteners
Illustrates a fastener that must be properly tightened with a torque wrench at this point in the rebuild. The torque specs are usually provided in the step.

Special Tool
Illustrates the use of a special tool that may be required or can make the job easier (caption with photo explains further).

Performance Tip
Indicates a procedure or modification that can improve performance. The step most often applies to high-performance or racing engines.

Critical Inspection
Indicates that a component must be inspected to ensure proper operation of the engine.

Precision Measurement
Illustrates a precision measurement or adjustment that is required at this point in the rebuild.

Professional Mechanic Tip
Illustrates a step in the rebuild that non-professionals may not know. It may illustrate a shortcut or a trick to improve reliability, prevent component damage, etc.

Documentation Required
Illustrates a point in the rebuild where the reader should write down a particular measurement, size, part number, etc. for later reference or photograph a part, area, or system of the vehicle for future reference.

Tech Tip
Tech Tips provide brief coverage of important subject matter that doesn't naturally fall into the text or step-by-step procedures of a chapter. Tech Tips contain valuable hints, important info, or outstanding products that professionals have discovered after years of work. These will add to your understanding of the process, and help you get the most power, economy, and reliability from your engine.

BEFORE YOU BEGIN

The iconic air-cooled VW engine is shown here in its long-block form.

The early Volkswagen power-plants were a marvel of engineering, which is why they have endured for more than 80 years. Developed by Ferdinand Porsche and his team in the late 1930s, the final design was rolled out in 1938. The engine was a 25-hp air-cooled engine that had a top speed of 100 km per hour (62 mph) and was capable of 32 mpg while traveling with two adults and three children.

A Brief History of the VW Flat 4

Very few Volkswagens were produced during World War II. Most of the assembly plants were in shambles, and none of the Allies wanted to take over producing the Beetle. They said they didn't see a market for such an ugly little car. The English decided they would take on the task of producing the car in Germany to give the German economy a boost.

Though not a hit when first introduced to the United States in 1949, Volkswagen knew it could sell the car to the automobile-thirsty American public. When Volkswagen started to gain popularity, a setback almost spelled disaster for the manufacturer. Early imports had mechanical issues, mainly with crankshafts and connecting rods breaking. Germany wasn't about to let some bad publicity about the reliability of their little car stop them, so in typical German fashion, the engineers overengineered the components in question. Throughout their production, components were continuously refined and upgraded.

Horsepower numbers continuously increased over the years. In 1954, the displacement was 1,192 cc and produced 36 hp. In 1961, the horsepower increased to 40 hp. Five years later, the horsepower jumped from 40 to 50 with the introduction of a new cylinder head design and displacement of 1,300 cc. In 1967, the displacement grew to 1,500 cc.

In 1968, Volkswagen had its best annual US sales year ever when it sold nearly 400,000 units at a base price of $1,699. In 1970, the final displacement increased from 1,500 to 1,600 cc, which increased horsepower to 57. The dual-port head was introduced in the Type 1 in 1971, and the horsepower topped out at 60. Finally, electronic fuel injection was introduced in 1975. After that, the engine remained relatively unchanged until the final VW Beetle was imported into the United States in 1979. There were more than 21 million air-cooled Volkswagens total built as late as 2003.

Engines

This publication covers the most-popular powerplant Volkswagen produced. These engines were installed in Type 1 Beetles, Karmann Ghias, Type 2 Buses (up to July 1971), Type 3 Squareback and Fastback, and the Type 181 Thing. Generally, every vehicle Volkswagen produced starting with the 1961 model year that used a Type 1–based engine will be covered within these pages.

Not only will this book cover the engines produced by Volkswagen but also the Type 1–based aftermarket engines. Did you know you can build a 100-percent brand-new air-cooled VW engine from aftermarket parts? This book will guide you with building those as well. With good rebuildable core engines becoming hard to find, this may be an option for you.

Type 1
In 1961, Volkswagen completely redesigned the Type 1 engine. Starting with a clean slate, very few components could be interchanged between the earlier 36-hp engine and the new 40-hp engine. Besides getting 4 hp more out of the same displacement (1,200 cc), many design elements were improved for

---

Engine or Motor? The Great Debate

You may have seen discussions on the internet regarding what people like to call the device that motivates their vehicle. Notice how I avoided saying "motor vehicle"?

According to Merriam-Webster:
engine: a machine for converting any of various forms of energy into mechanical force and motion; also : a mechanism or object that serves as an energy source.
motor: one that imparts motion; specifically : prime mover
2: any of various power units that develop energy or impart motion: as
a: a small compact engine
b: internal combustion engine; especially : a gasoline engine

Many people get caught up in the notion that engines use a fuel of some sort to convert energy into movement and that motors are generally reserved for electric-powered devices. That theory has a lot of holes in it when major manufacturers, such as General Motors, Ford Motor Company, BMW, and Harley-Davidson Motor Company, have the word *motor* right in their names.

When I change the oil in my vehicle, I use "motor" oil. It says so right on the bottle. And what about motorcycles; it wouldn't sound right calling them enginecycles. According to the definition, a motor is a type of engine, so I guess it really doesn't matter.

For the sake of continuity in this book, a Volkswagen engine will always be called an engine and never a motor. Even though, personally, I enjoy calling them motors. It just rolls off the tongue better. ∎

---

ease of manufacturing and increased reliability.

For example, the fuel pump was relocated to the right of the distributor. The generator stand was separated from the engine case. The cam follower and pushrod became two separate components; they previously had a cumbersome one-piece design. The valves were angled in the cylinder heads, increasing flow. A spiral groove was cut in the backside of the crank pulley to return oil back to the crankcase during operation. These improvements were only slightly refined for the entire run of this design.

During the four years the 40-hp engine was available in the United States, Volkswagen produced more than 3.5 million vehicles worldwide. The US Type 1 VW market ended

with the 1979 Beetle convertible, but this engine was produced by VW worldwide until 2003. It's no wonder many of these engines are still around today.

Displacement and Horsepower Changes
In 1966, the United States was just at the beginning of what is known today as the muscle car era. Nearly every automobile manufacturer was caught up in the frenzy, including Volkswagen. The Type 1 engine got a whopping 25-percent increase in horsepower for 1966, growing from 40 to 50 hp!

The increase was due to many factors. The displacement was increased from 1,200 to 1,300 cc and was noted by a "1300" badge on the decklid. This was one of the

---

Pre-1961 Engines and Practicality

Owners of earlier models with 25- and 36-hp engines might find this book useful, but those engines are not its main focus. These earlier engines have so little in common with the versions that are 40-hp and newer that it would be confusing to interject those differences within the context of building this more-popular version.

Parts are scarce for 25- and 36-hp engines and thus expensive for any pre-1961 VW engine. For this reason, it's very common to swap the more powerful and less expensive engines from newer cars into these earlier models. A 40-hp engine is a direct bolt-in conversion for these earlier models. Unless the owner is a purist and is looking for a historically correct restoration, this is the route generally taken. ∎

Engine Codes

This book will cover most versions of VW Type 1 engines after 1961 with the introduction of the 40 hp all the way through the fuel-injected 1,600-cc engines introduced in 1975. The following table provides the engine codes to identify a VW engine. ∎

Type 1			
Code	Year	Engine	Notes
1-	Pre-January 1956	25 hp/36 hp	1- prefix indicates Type 1 and is not part of the sequential engine number
2, 3, 4	1955–1965	1,200-cc 30-bhp DIN, 36-hp SAE	"A" series engine
5, 6, 7, 8, 9	1961–1965	1,200-cc 34-bhp DIN, 40-hp SAE	"VW D" added below serial after 9247364
D0, D1	1966–1985	1,200-cc 34-bhp DIN (German)	40-hp SAE for Mexico made 1,200 cc after January 1978
E0	1966–1970	1,300-cc 37-bhp DIN	Non-US M240 low compression
F0, F1, F2	1966–1970	1,300-cc 40-bhp DIN, 50-hp SAE	Only in 1966 for US, 1966–1970 elsewhere
H0, H1	1967–1970	1,500-cc 44-bhp DIN, 53-hp SAE	Only in 1967 for US, 1967–1970 elsewhere
H5	1968–1969	1,500-cc 44-bhp DIN, 53-hp SAE	M157 USA/Canada
B6	1970	1,600-cc dual relief, single port, 47-bhp DIN, 57-hp SAE	M157 USA/Canada
L0	1967–1970	1,500-cc 40-bhp DIN	Non-US M240 low compression. NOTE: all the following 1971-and-up engines are dual relief and dual port.
AB	1971–1973	1,300-cc 44-bhp DIN	Non-US
AC	1971–1972	1,300-cc 40-bhp DIN	Non-US M240 low octane
AE	1971	1,600-cc 50-bhp DIN, 60-hp SAE (gross)	US only
AE	1972–1973	1,600-cc 48-bhp DIN, 48-hp SAE (net)	US only with compression ratio reduced
AF	1971–1982?	1,600-cc 46-bhp DIN	Non-US M240 low octane
AH	1973–1974	1,600-cc 48-bhp DIN, 46-hp SAE (net)	US only
AK	1973	1,600-cc 48-bhp DIN, 46-hp SAE (net)	US only
AJ	1975–1979	1,600-cc 50-bhp DIN, 48-hp SAE (net)	Fuel injected
AR	1974–1975	1,300-cc 44-bhp DIN	Non-US
AS	1974–1979	1,600-cc 50-bhp DIN	Non-US
ACD	1992–2004	1,600-cc 46-bhp DIN (Mexico)	Fuel injected

few times Volkswagen ever bragged about engine size. The "1300" only lasted one year. The cylinder head was redesigned into what is known today as the single-port head.

By 1967, automobiles were becoming faster and more refined every year. Manufacturers tried to outdo each other in order to sell more vehicles than the previous year. Volkswagen was no different, and the horsepower increased again. Displacement increased from 1,300 to 1,500 cc, netting a 3-hp gain, and the clutch diameter increased from 180 to 200 mm to handle the increased power. The charging system finally joined the rest of the industry by becoming 12 volts instead of 6 volts.

The single-port head had its final run in 1970, aided by an increase in displacement from 1,500 to 1,600 cc. The 1,600-cc single-port engine became the favorite for many years for its mixture of old-world simplicity with more power.

A key year in the development of VW's engine program was 1971. This was the year the

Type 2			
Code	Year	Engine	Notes
20-	Pre-January 1956	25 hp/36 hp	20- prefix indicates Type 2 and is not part of the sequential engine number
2, 3, 4	1955–1960	1,200-cc 36-hp	Includes "Bastard 40 hp"
5, 6, 7, 8, 9	1961–1965	1,200-cc 40-hp	
O	1963–1965	1,500 cc	No cam bearings
H0	1966–1967	1,500 cc	
L0	1967	1,500 cc	M240 low compression
B0	1968–1970	1,600 cc	Non-US
B5	1968–1969	1,600-cc single relief	US only
B5	1970	1,600-cc dual relief	US only
C0	1968–1970	1,600-cc 44-bhp	Non-US, M240 low octane
AD	1971–1973	1,600-cc 50-bhp DIN	Non-US
AB	1971	1,300-cc 44-bhp DIN	Non-US, M252
AE	1971	1,600-cc 50-bhp DIN, 60-hp SAE (gross)	US only
AS	1974–1979	1,600-cc 50-bhp DIN	Non-US
CB	1972/1973	1,700 cc	Dual carburetor, manual transmission
CD	1973	1,700 cc	Dual carburetor, automatic transmission
AW	1973/1974	1,800 cc	Dual carburetor
AW	1975	1,800 cc	Fuel injected
AP	1974/1975	1,800 cc	European only
ED	1975	1,800 cc	
GD	1976/1977	2,000 cc	
CJ	1976–1979	2,000 cc	European only
GE	1978/1979	2,000 cc	
CU	1980–1983	2,000 cc	Vanagon style
CV	1980–1983	2,000 cc	Vanagon style

Type 3			
Code	Year	Engine	Notes
O	1961–1965	1,500 cc	
K0	1966–1972	1,500-cc 54 hp	
M0	1966–1972	1,500-cc 52 hp	M240 low compression
P0	1966/1967	1,600-cc 50 hp	M240 low compression
T0	1966–1973	1,600 cc	
U0	1968–1973	1,600 cc	Fuel injected for US
U0	1970	1,600 cc	Dual-relief case
U5	1971–1973	1,600-cc, 7.7:1 compression	US/Canada M239 California fuel injection
X	1972	1,600-cc, 7.3:1 compression	California

1,600-cc dual-port engine was introduced. It produced a neck-snapping 60 hp! That's 50-percent-more horsepower than the same basic design made 10 years prior. This version of the Type 1 engine remained relatively unchanged for the next three years. There were changes made to the charging system, replacing the generator with an alternator, adding a quieter muffler, and using different ignition timing that reduced the horsepower to 58 in 1973.

Beginning in 1975, all US Beetles and Super Beetles were fuel injected. They remained that way for the duration of their production. The Bosch L-Jetronic fuel injection reduced emissions enough to satisfy the US government while getting only slightly better fuel mileage. The horsepower numbers suffered in the balance, making only 50 hp.

Engine Identification

Identifying a Volkswagen engine is relatively easy. Along with the first letter or letters of the serial number, each engine provides clues to its pedigree with other factors. Knowing a few of these clues will come in handy when you are trying to decipher what you are looking at online or at a swap meet.

Maybe you are looking into purchasing a VW and you want to determine if it has the original engine. Spoiler alert! It probably doesn't! A little knowledge can save a bunch of time determining exactly what engine you are looking at.

Intake Manifold

A seasoned professional can easily ballpark the VW engine version by looking at the intake manifold. There are only three basic intake

The three different carbureted intake manifolds covered in this manual are: 40-hp 1961–1965 (left); 1,300–1,600 single-port head 1966–1970 (center); and 1,600 dual-port head 1971–1974 (right).

manifolds for the carbureted Type 1 engine: Type 1 Beetles with 40-hp from 1961–1965, single-port 1966–1970, and dual-port 1971–1974.

The easiest one to pick out is the dual-port manifold installed on 1,600-cc engines in 1971-and-newer engines. It has a three-piece design with a center section and aluminum castings that bolt to the heads. A dual-port manifold will always have a 1,600-cc engine. Remember that Volkswagen did make a 1,600-cc single-port engine; so if someone says it's a 1,600, don't assume it's a dual-port engine.

If the engine has a three-piece intake manifold, it is a 1,600-cc engine. Volkswagen made a 1,600-cc single-port manifold in 1970, and it can be hard to distinguish this from a 1,300-cc (1966) or a 1,500-cc (1967–1969) just by looking at it. The serial number on the case of a 1,600-cc single-port will start with the letter *B*.

Single-port engines look very similar to the earlier 40-hp to the untrained eye. The intake manifolds

look similar but aren't the same. The easiest way to tell them apart is to look at where the manifold bolts to the head. The 40-hp and single-port share the same size hardware (6-mm studs) where they bolt to the head, but the studs on the 40-hp point straight down. The single-port manifold studs are at an angle.

Also early 40-hp cases did not have cam bearings. The cam rode directly in the case much like a cam in a motorcycle. Late-1965 40-hp cases started to get cam bearings, but there is no guarantee what cases got the bearings because the factory was already changing over to the new 1,300 engine.

The single-port engines are almost impossible to tell apart without looking at the beginning of the serial number. They came in three different displacements: 1,300, 1,500, and 1,600 cc. The 1,300-cc engines were the oddest of them all because they only came in the 1966 Beetle in the United States. A mixture of old and new parts, they really aren't the most desirable engine to have unless you are restoring a 1966 Beetle to absolute 100-percent stock condition. Then you might be able to justify the more expensive parts it will take to rebuild it due to the rarity of the parts on a one-year-only build.

If you have a single-port-engine vehicle (1966–1970) and want to keep it looking period correct but have a little more horsepower, building a 1,600-cc single-port is the way to go. Starting with a 1,300-cc will not be cost effective due to the amount of machine work that will be needed to the engine case and cylinder heads for the installation of the larger pistons and cylinders. But the 1,500 cc available from 1967 through 1969 can easily be converted to 1,600 cc by simply changing out the pistons and cylinders. No machine work is necessary due to the outer dimensions of the cylinders being the same. You will get the benefit of more displacement and horsepower with minimal work and expense.

Here is an example of the serial number of a factory manufactured engine case. The previous serial number was machined off with the exception of the A *that designated it as a dual oil pressure relief case. The VW re-manufacturing symbol was stamped in and then the letters AK. The area next to that might have the serial number that was being replaced stamped in, but it rarely does.*

Serial Number

The only way to be sure of what you are looking at is to read the serial number stamped into the engine case. On all Type 1–based engine cases, the serial number is located just beneath the generator stand and just above the dipstick. In the United States, the serial numbers on early 40-hp engines didn't begin with a letter; instead, the numbers started with a 5, 6, 7, 8, or 9. Worldwide the letter *D* designated the 40-hp engine cases, so the case may or may not have a letter *D* stamped near the serial number.

In 1966, Volkswagen started designating engine models by adding a prefix of letter or letters to the serial number. All of the single-port engines have just one letter, starting with the letter *E* for the first 1,300-cc single-port engines.

With the introduction of the 1,600-cc dual-port engine in 1971 came a new numbering system. All of the 1,600 Type 1–based engines had two letters to start the serial stamped on the case. The first letter is always an *A*. It's super easy to remember: no letters means a 40 hp, one letter is single-port, two letters that start with *A* is a 1,600-cc dual-port.

Other Indicators

There are additional pointers that are helpful when trying to determine what vintage engine is in your car. The fuel-injected versions introduced in 1975 are easily identified by the lack of any mechanical fuel pump machining done to the case just left of the alternator stand.

Say the engine is not installed in a vehicle and you can't find the serial number in the usual spot under the generator stand area. It could be out of a Type 3 (fastback or squareback). The absence of a dipstick will be your first clue.

Type 3 engines will have the serial number on top of the case right behind the fuel pump. Right along where the two engine case halves come together. No numbers there either? More than likely it's a factory replacement case or even an aftermarket replacement case.

Years ago, a replacement case could be purchased directly from a VW dealer. Today, aftermarket manufacturers have stepped up and now produce OEM-quality replacement engine cases. To keep costs down, these cases are universal, meaning they are made to fit almost every configuration of Type 1, Type 2, and Type 3 engine. The only downside is they are only available in the 1,600-cc bore size: 85.5 mm.

Is This Engine Worth Rebuilding?

Many factors will determine if it is cost effective to rebuild a certain engine. Every engine *could* be restored but not every engine *should* be restored unless it holds special sentimental value or is super rare. Your engine may be beyond repair and may only be good for a few usable parts.

Let's start with a visual inspection. Does it have a giant hole in the top of the case? If so, more than likely, it threw a rod and the entire lower end is beyond repair.

Has it been left outside in the elements uncovered? Is it lying directly on the dirt floor? Mother Nature is the number one killer of stored vehicles and their components.

Read the dipstick. Does it have any oil? What does it look like? Black is actually okay. What you absolutely do not want to see is water. If the oil is milky like coffee with a lot of cream or even chocolate milk or has any presence of water at all, it might not be a good candidate for a rebuild. What happens is the water, being thinner than oil, settles down at the bottom of the sump. Water and magnesium do not play well together. Corrosion happens very quickly and will destroy the engine case from the inside out. Corroded magnesium is impossible to weld cleanly, and corrosion around the sump plate can render an engine case unusable.

If the engine is installed in a car and it runs, more than likely it will be rebuildable. Even if it runs poorly; is missing, chugging, or knocking; is running on three (or fewer) cylinders; is leaking oil from every corner; etc., it more than likely will be just fine once rebuilt.

If it doesn't run, the first step is to determine if the engine is locked up. Grab the fan belt with both hands and try to rotate the engine one way or the other. No fan belt? Try turning the pulley or even the flywheel. If it rotates easily, it's good. If it's difficult to move but it does move some, it's likely still rebuildable.

If the movement stops abruptly in either direction, more than likely the lower rotating assembly is unserviceable and the engine would not be a good candidate for a rebuild. If it rotates slightly but binds up at any point and doesn't spin completely around, then it's usually just the piston rings dragging on rust in the cylinder bores. If this is the case, the engine is more than likely serviceable but it may take some work to get apart.

The worst-case scenario is that the engine is locked up tight and doesn't move in either direction. Try pushing and pulling on the lower pulley. Do you have a little movement? That's the glimmer of hope you are looking for. It's going to be a bear to get apart, but it may be savable. The value of this condition engine is very low, like almost free. Take this into account. You may want to look into another prospect.

Planning

Starting any project always requires three things: knowledge, money, and time. With this book will come knowledge. Rebuilding the engine yourself will save money. The last thing is time. How long is this going to take? A big factor with time is your mechanical ability, but some factors will be out of your control. For example, the machine shop might have a backlog or your parts might not arrive in a timely manner. Preplanning will speed things up and keep you on track.

Plan for engine teardown to take 8 to 10 hours. An entire weekend should be enough time to get the engine completely apart with the parts sorted. Organize the parts into three piles: parts to clean up, parts to throw away, and parts to take to the machine shop. While parts are at the machine shop, the rest can be cleaned, inspected, prepped, and painted.

The machine shop will determine what bearings you need and if anything is beyond servicing. Once you get that information, you can place an order for the parts you need. Don't be surprised if you find things later that should be replaced as well. This process may take a few weeks, depending on the machine shop time and how long it takes to get all your parts.

Assembling a long-block will take about 10 to 12 hours, which is slightly longer than the teardown process. It will take another 4 to 6 hours to get the engine to the point it can be started. This includes installing the cooling tin, fan shroud, charging system, ignition, and intake and exhaust systems.

A realistic time frame is three to four weeks total, which really isn't too bad. In fact, that's quicker than some shops that have a waiting list for complete engines.

Work Area

Most professional shops are divided into two areas: one to tear down engines and another for assembling engines. The reason for that is to keep the mess separate from the clean parts. This can be achieved by the home builder by doing the teardown in a different area from where the assembly is going to occur. The driveway or garage stall works just fine.

Disassembly can be quite messy and can occur just about anywhere, but a clean, dedicated area to assemble your engine is a must for a

Tech Tip

To take apart a serviceable engine, remove the spark plugs and spray penetrating oil or WD-40 down the holes. You will save yourself a lot of frustration by doing this well before you start the disassembly process. ∎

successful build. Most home workshops are small, but as long as the previous mess is cleaned up and you start fresh with a clear workbench and all the tools put away, you're off on the right foot. An extra table close by helps as well to lay out parts and stay organized.

Lighting will also play a key role in how well your build turns out. There's a reason the operating rooms are the best lit rooms in the hospital.

Your build may get put on hold for any amount of reasons. Be sure this area won't be disturbed if that happens.

Cleaning Area

Cleaning parts is not the most popular aspect of engine building, but it is a very important one. Mineral spirits can be used as a solvent as well as some water-based parts cleaning solutions available at most tractor supply stores.

If you don't have a parts washer but have access to one, by all means use that favor and clean your parts. If that's not an option, a small bench top or floor model can be purchased for less than $100. Remember to place the cleaning tank in a well-ventilated area and add an exhaust fan if necessary. As an additional safety measure, keep a dry-chemical fire extinguisher handy.

To save some time and get a more thorough cleaning job, have the machine shop clean your engine components. For a fee, they will run your parts through an industrial parts washer or hot tank. Just be careful and keep track of what you bring there so you are certain to get it all back. Take a photo if necessary.

Tools

Volkswagen designed its engines early on with the idea that the average mechanic should be able to do most of the routine maintenance and even some of the more involved repairs right in his or her own garage. With this in mind, it limited the fastener sizes to just a few. VW mechanics joke they can take apart an entire Beetle with five wrenches: 8, 10, 13, 17, and 19 mm.

The internal components of the engine will add a couple more sizes to the mix, but if you own a decent set of metric tools you will be fine. You will also need combination wrenches; 1/4-, 3/8-, and 1/2-inch drive sockets; and ratchets to get the majority of the work done.

The majority of the fasteners that need to be torqued can be handled with a 3/8-inch drive torque wrench. Don't rely on the cheap beam-type torque wrench lying in the bottom drawer of the toolbox. If you need to buy one, don't get a cheap one. Buy a name-brand wrench that has a micrometer type adjustment and clicks when the set torque is reached. The latest electronic ones have LED lights that tell you when you are getting close to the correct torque then beep when the set torque is reached.

To torque the flywheel to the crankshaft, you will need a serious torque wrench that can measure at least 253 ft-lbs. The casual mechanic will find it hard to justify buying a torque wrench that can handle that much torque just to tighten one fastener. A 3/4-inch drive torque wrench can easily cost hundreds of dollars but can be rented inexpensively, and some major auto parts retailers will even lend you one for free with a deposit.

An air compressor or access to compressed air will be a must to clean components and blow out passages as well as operate air tools if you own some. Tearing down an engine will go much quicker with air tools or cordless impact tools if you have access to those.

Here is a list of tools you will need other than your basic hand tools.

- 3/8-inch drive torque wrench—10–75 ft-lbs range
- 3/4-inch drive torque wrench—50–300 ft-lbs range

You will need two torque wrenches to rebuild an entire VW engine. Most fasteners only require a 3/8-inch drive, such as the one shown. The "clicker" type is recommended over the beam type due to their increased accuracy. You will also need a larger one that can torque the flywheel to 253 ft-lbs.

A professional piston ring compressor is well worth the money. This ring compressor by K-D tools is made in the United States and built to last. The pliers (model 850H) and ring (model 850-BC) have a 3 3/8 to 3 5/8-inch range.

In order to set and check the crankshaft end play, a dial indicator and magnetic base are a must. This Starrett indicator and Cullen magnetic base are examples of what you need but are, by no means, the caliber of tools necessary. An indicator and magnetic base from Harbor Freight Tools for around $29 will get the job done just the same.

- Piston ring compressor
- Dial indicator with magnetic base
- Digital calipers
- Professional lock ring pliers
- 1 $\frac{7}{16}$-inch (36 mm) 3/4-inch drive socket
- 3/4-inch drive breaker bar

Removing the clip that holds the crankshaft gears on can be challenging, to say the least. A good pair of lock ring pliers are about the only way that clip is coming off. The Craftsman models (47386 and 46948) shown here work quite well. Wilde Tool lock ring pliers (model G409.B) are made in the United States and only cost $20.

If you have one of these, you already know what a joy it is when this fine device is in your hand. This particular Craftsman model (41588) is perfect for prying things apart or making sure that things you were thinking about replacing are now on the "need to replace" list.

In order to heat up the crankshaft gears to slide them onto the crank, you will need a source of heat. A single burner hot plate can be purchased for less than $20 and works quite well. It sure beats getting the kitchen all smoky heating up gears on the stove and taking heat from your better half.

- Screwdriver
- Hot plate or quick access to a stove
- Parts washer or access to one
- Air compressor and blow gun

Specialty Tools

As you can imagine, there will be some specialty tools needed to get the job done. These tools will make difficult jobs easy. In some cases, they will be the only way you can assemble an engine correctly. All of these tools are VW specific.

This basic engine stand is one you may already have or can easily buy online or at Harbor Freight Tools. It's rated at 1,000 pounds and costs $60. The engine stand yoke is specifically designed for a VW engine from Vintage World Tools ($91). The drip pan is actually a restaurant 19-gauge aluminum sheet pan measuring 18x26 inches (Thunder Group ALSP1826) and costs less than $7.

A floor model engine stand (EMPI 5007) is more mobile and can be disassembled and stored for a less permanent solution. (Photo Courtesy EMPI Inc.)

A flywheel lock holds the flywheel to the case so the gland nut can be removed and installed. This lock (EMPI 5003) fits both 6V and 12V flywheels. (Photo Courtesy EMPI Inc.)

A clutch pilot tool can be purchased rather inexpensively, but those are usually made of plastic and aren't very precise. Shown are two inexpensive alternatives. The top one is made of aluminum and was turned up on a small lathe. The smallest diameter must slip into the gland nut pilot bearing and the next step must be a slip fit into the clutch disc. The lower one is the input shaft from any VW Type 1 transmission. Most local VW shops will gladly sell you one inexpensively.

A VW air-cooled-specific engine stand is a must to rebuild an air-cooled VW engine. A bench-mounted one (EMPI 5001) is most economical and works if your bench can hold the weight. (Photo Courtesy EMPI Inc.)

- Engine stand
- Flywheel lock
- Clutch pilot tool
- Crank gear puller
- Lifter holders

- Crankshaft vise mount
- Flywheel holder "Helping Hand"
- Crankshaft hand crank

- Oil pressure relief plug tool
- Bent 13-mm box wrench

DIY Tools

Sometimes the simplest things can make life so much easier. Planning ahead and making sure you have the right equipment can make all the difference. These tools are ones you can make yourself to help with the engine rebuild.

If your buddy or the local machine shop doesn't have a VW-specific crankshaft gear puller, your only other option is to buy one. Most shops charge $15 to $20 to remove the gears. This puller (EMPI 5714) costs $80 to $90. (Photo Courtesy EMPI Inc.)

These lifter holders are clips made to hold the lifters into the 3-4 side of the case while you set it down on the other half. Most people use grease or assembly lube to hold them in, but lifters may slide out of their bores and down into the other half. If you run into issues causing you to start over, it takes longer than it should. (Photo Courtesy EMPI Inc.)

This flywheel holder is made of four items and is a sure-fire way of removing a flywheel gland nut. Gather a 1 ⁷⁄₁₆-inch (36-mm) 3/4-inch drive socket, a 3/4-inch drive breaker bar, a length of pipe that will fit over the breaker bar, and a length of heavy angle iron with a couple holes drilled in it to fasten it to the flywheel. Drill one 3/8-inch hole near the end and another 7 ⁷⁄₁₆-inch from there. Length is unimportant but remember leverage is your friend.

It may not seem like a valuable tool, but you'll feel like a professional cranking over your fresh rebuild with your very own hand crank. To make your own, bend a length of 3/8-inch rod and weld it to a crank pulley bolt. You can use an old 10-mm head stud if that's what you have handy.

Shop Supplies

These items may seem obvious, but a list is a good way to make sure you are prepared.

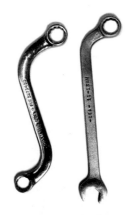

The pressure relief screws can seem impossible to remove. That's where this handmade tool will come in handy. You'll need a 3/4-inch bolt that is 4 to 5 inches long. Grind the end until it fits snugly in the screw. It has a slight radius along the blade, as that is how it is in the screw head. Give this tool a couple sharp blows with a hammer to shock the relief plug loose.

A curved 13-mm wrench can be a time-saver when removing and installing a VW carburetor. You can pay for a fancy Snap-On version or make your own out of a generic 13-mm wrench and a propane torch.

An easy solution to holding the crankshaft safely is this vise mount made from an old flywheel gland nut and a short piece of angle iron. Cut the angle iron the width of your vise jaw and weld the nut to it. By screwing the holder into the crank and then clamping it into the vise, you have easy access to all the components attached to the crankshaft.

Red Line makes a great engine assembly lube that works well for cam lobes, lifters, and lifter bores. All the direct, metal-to-metal components can benefit from this product.

General assembly lube that is easily applied can be made by mixing equal parts STP oil treatment and 20W50 motor oil. Sometimes it's a little hard to get pumping out of the oil can, but it can be applied with one hand and without getting your hands all sticky.

Permatex aviation sealant has been around forever and is about the closest product available that mimics the original sealant used by the factory. It is primarily used to seal two machined surfaces without gaskets, such as the two engine case halves. It also works well for tacking gaskets in place during assembly.

- Assembly lube
- Permatex aviation sealer
- STP oil treatment
- Castrol GTX 20W50 motor oil
- Silicone gasket maker

RTV or silicone gasket sealer should be used very sparingly on a VW air-cooled build. Never use it to seal the two case halves together. Common uses are valve cover gaskets and pushrod tube seals. This Permatex product has great oil resistance and seals off those areas well.

Before you get too far into the process, make sure you have an ample supply of brake cleaner and paper towels. A box of lint-free shop towels is a good alternative paper towel for wiping off critical components, such as bearing journals, prior to assembly.

- Brake cleaner
- Paper towels
- Various containers

Our Engine

The focus in this book will be on the most popular and most desirable air-cooled flat 4-cylinder engine that Volkswagen ever produced. It is no coincidence that the most popular engine is also the most powerful stock 1,600-cc air-cooled engine that Volkswagen ever produced for the American public. With its 60-hp engine and upgraded cooling and oiling systems, the engine installed in 1971–1972 Beetles, Super Beetles, and Ghias will be the engine we will use for the detailed disassembly and reconstruction.

Simple and inexpensive items work great for organizing parts. Muffin tins are great for sorting hardware, and cake and bread tins help with larger items. All of these items can be found at your local thrift store, at a rummage sale, or possibly in your own kitchen. Maybe your wife would like new stuff and your anniversary is coming up. Problem solved.

Volkswagen produced more than 700,000 units with this powerplant during those two model years. The serial number designation is AE. It has a 1,600-cc engine with dual-port heads, a doghouse-style oil cooler, a dual oil pressure relief case, a wider engine cooling fan, and a generator charging system.

Our particular engine was previously installed in a 1970 Beetle that was converted to a Baja Bug. As in most cases where the engine isn't original to the car, the mileage is unknown. We didn't even try to start it because we knew it was hard to turn over by putting a wrench on the crank pulley bolt. Even with the spark plugs removed and oil squirted into the cylinders, it was difficult to turn over. Since there weren't any obvious signs of complete destruction, we decided this would be a good candidate for a rebuild. Fingers crossed.

This engine was plucked from a 1970 Beetle that was converted into a Baja Bug. Much of the engine tin was already missing due to the fact that it was installed in an open air–type vehicle. The distributor had been changed out for the more popular Bosch "009" mechanical advance distributor.

Here you can see the improved cooling system that was introduced in 1971. Commonly known as the doghouse cooling system, it was a huge improvement. The oil cooler moved from inside the fan shroud to its own housing behind the fan shroud. The oil cooler increased in size and was now made of aluminum.

This is the serial number for our engine. Notice the location is right under the generator stand and above the dipstick. The number AE699256 designates it as a 1971 1,600-cc dual-port engine.

ENGINE DISASSEMBLY

It is human nature to be curious and wonder about the past a certain object has had. The old adage "if these walls could talk" is very fitting to the life of an engine and what it has endured over the years. With all the millions of Volkswagen engines still in service, the possibility of your engine having as many lives as a cat is not out of the question. Telltale signs of engines being rebuilt three, four, five, or more times may appear during the disassembly process.

Spotting a Rebuilt Engine

Most rebuilt engines are very obvious. The most obvious sign is a paint job. Volkswagen never painted its engines from the factory. Sure, the valve covers and other sheet metal was painted a semigloss black, but the engine case and cylinders were never painted. They had a coating that preserved them and prevented them from oxidizing for a period. The coating had a slight maple syrup color to it and didn't last very long. Though a paint job doesn't always constitute a rebuilt engine, a very good paint job is hard to achieve without cleaning all the grease and grime out of every nook and cranny and getting the paint to look nice.

Another sign that the engine has been rebuilt is random hand-stamped letters or numbers other than the factory serial number. These are usually stamped near the generator stand area. Don't be discouraged if your engine has been rebuilt by a less-than-reputable rebuilder. As long as the majority of the components are intact, everything should be fine.

Be Patient

Don't get frustrated during the disassembly process when fasteners may seem to be welded together. Remember, this piece of equipment might not have had a wrench put to it in more than half a century. Be patient and use the right tools.

Solvent-based rust penetrants, such as Liquid Wrench, WD-40, PB Blaster, etc., can help loosen things up. Apply them days in advance if you can, and don't be stingy.

Apply heat if necessary to help loosen stuck fasteners. MAPP gas is a great, inexpensive source of intense heat. It works when a standard propane torch falls flat.

Tech Tip

Tearing down an engine "just for fun" is a good learning experience. Perhaps you came across a two-for-one deal while shopping for an engine to rebuild or someone is willing to give you a free engine to get it out of their way. Older 6-volt engines and partially disassembled engines are great candidates for an afternoon of fun and education. Whatever the case, it's a great way to see what you are getting into before actually tearing down the engine you are going to rebuild. ■

If you are putting enough force on a fastener that you think it's going to snap off, more than likely it will. There is nothing worse than a snapped off stud. Broken exhaust studs are super common and there is no reason for it to happen. An easy way to remove the nut is to sacrifice it. By that I mean to use the sharpest chisel you own and chisel the nut in half. Nuts are cheap, but broken studs are not.

The small 6-mm screws that hold on the sheet metal are commonly

A Word About Safety

While you are swinging hammers and lighting things on fire, please use your head and be safe. Take your time and think about the what-ifs. Safety is not only the smart thing to do but also the most economical. A trip to the emergency room to get stitches or to get something removed from your eye can be very expensive and really throw a wrench in the budget and the timeline.

After you have an accident you always have the feeling of remorse. You instantly regret not being more careful or not taking the necessary precautions. So wear those safety glasses and put on those gloves. Get help if you need it and take a break if you need one. Nothing is more important than your well-being, and nothing slams the brakes on a project faster than a needless injury. ∎

frozen as well. Hitting the top of the head of the screw to "shock" it loose works great. It's worth a shot before pulling out the big guns (i.e., the torch). If hitting it doesn't work, use an acetylene torch to heat the area around the screw. It will expand and break the fastener loose.

Get Organized

You've gathered all the tools and equipment necessary to start tearing down the engine. Before you begin, you will also want to gather items to keep your disassembly organized. The last thing you need is to start misplacing items from the very start.

Plastic bins are great for staying organized and hold up much better than cardboard boxes that just soak up oil and fall apart. The bins are available at most warehouse and home-improvement stores. Get enough bins to separate parts into three categories: parts to clean up, parts to take to the machine shop, and parts to replace. The parts you need to clean up can be put into a few bins. Most of these things will be put through the parts washer and inspected later. The parts you are going to replace can be recycled or sold outright. No need to clean them for this project.

Get out zippered bags and permanent markers. Put specific hardware in the bags and write on the bag in a language you can understand. Don't get caught up in the correct terminology. The only person who has to understand what these things are and what they are used for is you. Putting smaller items to replace in metal bread tins or old-school coffee cans will keep everything in one place.

Don't throw anything away until you are positive you no longer need it. Even a junk part that will only be used for reference is worth saving.

Taking pictures is a great way to reference parts and assemblies later in the process. You can take photos with your smart phone or a digital camera of things while they are still together the correct way, as you disassemble, and once they are in parts. Keep the photos organized for referencing later. When in doubt, take a picture. It's easy and free.

Engine Disassembly

The main disassembly process will be exactly the same whether your engine is a 40-hp model from 1961 or a 1,600-cc fuel-injected model from 1979. The long-block will have nearly the same amount of parts and come apart in the same order. Let's go step-by-step through the entire teardown process.

Disassembly

1. Remove the Gland Nut

Removing the flywheel can be a chore. It's easiest to remove it before installing the yoke for the engine stand. The flywheel is held to the end of the crankshaft by one fastener called the gland nut, which is actually a big bolt torqued to 253 ft-lbs. Install the Helping Hand flywheel holder you made out of angle iron. Loosen the nut with a $1^7/_{16}$ (36-mm) socket and 3/4-inch drive breaker bar. A piece of pipe over the breaker bar will provide more leverage. Do not attempt to break the gland nut loose using a 1/2-inch drive socket and breaker bar. You will find the breaking point of these tools very quickly.

1. Remove the Gland Nut
CONTINUED

Shown here is the gland nut we just took off compared to a new one. You can tell someone had previously taken it off using the wrong tools (perhaps an adjustable wrench from the looks of the rounded off corners). This will need to be replaced with a new gland nut.

2. Remove the Flywheel

Use two long pry bars to wiggle the flywheel off the dowel pins. It shouldn't take much effort. Be careful not to let the flywheel fall to the ground. Once the flywheel was removed, it was obvious why the gland nut was removed once before. The factory rear main seal (which was black) had been changed out for a new modern rear main seal (orange in color and made of a better silicone-based material).

3. Bolt the Yoke to the Case

Attach the engine stand yoke to the back of the engine case. Notice that it is attached to the left or 3-4 side of the engine only. This is so when the engine case is split apart, one half will stay connected to the stand while you remove the other half. The yoke will not fit straddling both halves.

4. Lift the Engine into the Stand

Slide the yoke into the stand. It weighs roughly 200 pounds, so don't try to lift this alone. Get some help; better safe than sorry. Now is a good time to start draining the oil. Most sump plates will have a drain plug, but later models eliminated the plug, which forces you to loosen the sump plate at its six studs to get it draining.

5. Remove the Carburetor

Using a bent or curved 13-mm wrench makes removing the carburetor so much easier. Your carburetor may be full of fuel, so it is wise to empty it by turning it upside down and letting the fuel pour out the vent tube once removed. Note this is a stock Solex 34PICT carburetor and the vacuum lines have been plugged due to the vacuum advance distributor being replaced with a mechanical advance distributor.

6. Remove the Fan Shroud

Next to come off is the fan shroud and generator assembly. First, loosen the clamp holding the generator to the stand using a 13-mm wrench and a 13-mm socket. Slide it back toward the fan. Next, loosen the three bolts attaching the fan shroud to the tin work using a 10-mm wrench. There is one by each intake end casting and one by the oil cooler. Lift straight up off the oil cooler.

7. Remove the Intake Manifold

Loosen all four clamps that hold the sections of the intake manifold assembly together using a #2 Phillips screwdriver. Behind the fuel pump is a nut holding the center manifold to the case, remove it with a 13-mm socket. Next, remove the two nuts holding the end casting on the 3-4 side head with a 13-mm wrench. Pull the center section and 3-4 casting off as one piece. Finally, remove the casting on the 1-2 head using a 13-mm wrench.

8. Remove the Cooling Tin

This is what it should look like with all the cooling tin removed. All the sheet metal is attached with 6-mm flathead screws. Depending on how complete the engine you have is or what type of vehicle it was previously installed in will determine what sheet metal it has and what it will need when reassembling. Some of this hardware can be difficult to remove without sheering the heads off. Try to rap them with a ball-peen hammer to shock them loose and save you from having to drill them out and reestablish some threads in their place. Be careful using a locking pliers to loosen them. Penetrating oil and some patience would be a better route.

9. Remove the Crank Pulley

Here is a tried-and-true way of removing the crank pulley with minimal damage. Make sure the crank bolt was removed using a 30-mm socket and all the cylinder tin is removed, then tap the backside of the pulley closest to the center as possible with a large screwdriver and a hammer. It doesn't take much force and it should walk right off. Crank pulley pullers tend to distort and permanently bend the thin metal pulley.

Even if this engine ran, it wouldn't have lasted very long without severely overheating and self-destructing. The cooling fins are packed tight with oily crud, rat droppings, and other debris. Stored cars are magnets for furry animals to make homes and start a family.

10. Remove the Generator Stand

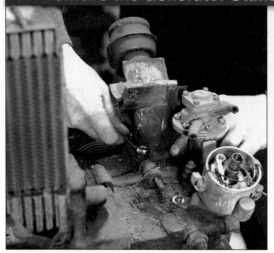

Remove the four nuts that hold the generator stand to the case with a 13-mm wrench. The baffle underneath the stand can be pulled off the studs. The one nut holding the distributor clamp to the case can come off as well.

11. Remove the Oil Cooler

Remove the three oil cooler nuts with a 13-mm wrench. There will be two nuts underneath by the 3-4 barrels and one up top toward the split in the case.

12. Remove the Sump Plate

By now, most of the oil has drained from the case and the entire engine can be flipped upside down. Remove the sump plate by removing the six nuts with a 10-mm socket. Pry off the sump plate with a screwdriver.

13. Remove the Oil Strainer

The oil strainer can be pried off the bottom of the case. It was intended to be serviceable by cleaning out and reinstalling it. This one is beyond that point and will be replaced.

Just look at that sludge! This engine was definitely neglected. Though obviously bad, it could be much worse. There are no signs of water damage; that thick, protective coating of goo preserved the sump area well.

14. Remove the Oil Pressure Relief Plungers

Special Tool

Use the special tool you made to remove the plugs. A couple quick blows directly on the head of the tool with a ball-peen hammer will shock them loose. The plungers should come out with a magnet.

Save Money 💲

💲 If the plunger is stuck at the bottom of the bore, an easy way to remove them is to start threading a 1/2-20 tap into the backside of the plunger. It works every time without buying a special plunger pulling tool. This is the same tap we use to install case savers.

15. Remove the Valve Cover

Rotate the engine 90 degrees in the stand. With a large screwdriver, pry the valve cover bail down. Remove the bail by pulling the ends out of the holes on either side of the head. If the valve cover is stuck, wedge the blade of the screwdriver between the boss for the end of the bail and the edge of the valve cover and pry up.

16. Remove the Rocker Assembly

Remove the rocker assembly by removing the two nuts with a 13-mm ratchet. The spark plugs can come out at this time as well. Note that the rockers aren't touching a couple of the valves, which is not a good sign.

17. Remove the Pushrods

Pull out all four pushrods. Put them directly in a large plastic cup to catch the oil draining out of them. Rotate the entire engine 180 degrees in the stand and repeat steps 15, 16, and 17 to remove the valve cover, rocker assembly, and four pushrods from the other side.

18. Remove the Head Nuts

The heads come off at this point. They are held on with special nuts and washers. Remove the eight nuts using a 15-mm socket and impact wrench if you have one. Start at the ends and work your way to the center when taking the nuts off.

Don't be alarmed if the nut gets stuck to the stud and pulls the entire stud out of the engine case. Eventually, all the head studs will have to be pulled out and case savers will be installed.

19. Remove the Head

Once all the nuts are removed, it's a matter of pulling the head straight up off the barrels. This may take some gentle persuasion with a dead blow hammer. Be careful not to hit any of the cooling fins on the head, which are cast aluminum and rather fragile.

Sometimes the barrels want to stay with the head while you pull them up over the studs and that's fine. Just be careful not to drop a barrel on your foot. The pushrod tubes will either stick to the head or simply fall out. They will be replaced, so discard them.

19. Remove the Head CONTINUED

Believe it or not, this head is probably salvageable with stuck valves and all. More concerning is the broken exhaust studs. A multitude of minor issues can render this head a bad candidate for a rebuild simply due to the cost of repairs, which may exceed the price of a replacement head. Closer inspection will determine if that is the case.

20. Remove the Deflector Tin

Use a screwdriver to pry off the lower cylinder deflector tin. All of the buildup on this important cooling component can lead to a hot spot and barrel distortion from uneven cooling.

21. Remove the Pistons

All VW air-cooled engines have floating wrist pins held in by clips. The clips on either side of the pin can easily be removed with needle-nose pliers. Using a 3-inch 3/8-drive extension pushed into a 6-inch 3/8-drive extension makes a handy tool to tap out the pins from the pistons. The crank pulley was partially reinstalled to rotate the pistons up to a point that the pins could be driven out.

22. Repeat Steps 18 to 21

Rotate the engine 180 degrees on the engine stand and repeat steps 18 through 21. At this point, you should see a short-block minus the flywheel (shown).

23. Remove the Oil Pump Cover

A cordless impact and a 13-mm socket make short work of removing these four nuts. Note the nuts have a special red plastic sealing surface and should be reused if possible.

24. Remove the Oil Pump Gears

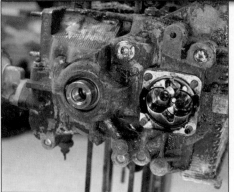

Once the two oil pump gears are removed, a special VW oil pump puller could be used to remove the gears. Occasionally, the pump is stuck in the case so badly that the puller will bend and even ruin the pump body before it comes out. The pullers have been found to be less than effective and at times do more harm than good.

25. Remove the Oil Pump Studs

A more-effective oil pump removal procedure does not require any special tools. Simply double nut two studs on one case half and remove two of the four studs. When you split the case, the pump will fall out.

26. Remove the Perimeter Nuts

Using a cordless impact, an extension, and a 13-mm socket, zip off the 14 nuts holding the two halves together. Three of these will be actual nuts and bolts; the other 11 are studs, one of which is the stud for the intake manifold center section. Be careful not to miss any. Some may be buried under dirt and sludge.

27. Remove the Main Bearing Nuts

The only thing keeping the two engine case halves bolted together now are the six main bearing fasteners. Use a 17-mm socket to remove them, starting with the end ones and working toward the center.

28. Pry the Case Apart

At this point, the case halves are ready to be pried apart. Double-check that all of the nuts are removed. Two 8-mm dowel pins locate the case halves together: one is at the top of the case toward the flange that mounts the case to the transmission and the other is right below the oil pump opening. Tap a thin screwdriver into the split line at the top of the case where it bolts to the transmission to get the two halves to split apart.

29. Tap the Case Apart

Knock the case half that sticks out near the oil pump opening with a dead blow hammer. This should make the two halves come apart evenly. Some gentle prying at these two points with a screwdriver should get them past the ends of the dowels.

30. Lift the Case Half Up

Grab the 1-2 case half by the head studs and slowly pull straight up. Don't be alarmed by the clunking noises. Those are the four lifters falling out of their bores and landing in the other case half. The engine case divorce is complete.

31. Remove the Seal and Shims

Discard the rear main seal and find three shims behind it. These shims of varying thicknesses will determine the crankshaft end play. Many factors go into what determines the correct thickness of shims to obtain the correct amount of end play.

32. Remove 3-4 Half Internals

No tools are necessary to remove the oil pump, cam, and lifters. Stick a finger in each end of the crankshaft and lift that out as well.

33. Remove the Key, the Oil Slinger, and the Nose Main Bearing

Special Tool

Screw your homemade crankshaft holder into the flywheel end of the crank and clamp the holder into a vise. Using a screwdriver and a hammer, tap out the pulley key. Remove the oil slinger and nose main bearing.

---

34. Remove the Crank Clip

With professional lock ring pliers, remove the clip above the gears. This will be difficult and nearly impossible to do without a good set of pliers.

Special Tool

35. Remove the Crank Gears

Oil the threads of the puller and place the U-shaped end underneath the cam gear. With a 3/4-inch socket and an impact gun, drive the puller into the end of the crankshaft. This will force the gears off. Don't be alarmed if this takes more effort than you think it should. The gears are an inference fit and have been on for decades.

36. Remove the Rod Nuts

Use a 14-mm socket and an impact driver to remove the nuts from the connecting rods.

37. Remove the Connecting Rods

Tap on the ends of the studs with a dead blow hammer to break the two parts loose. Keep the cap with the rod it came off; they are all numbered from the factory so you can pair them up correctly.

38. Remove the Fuel Pump and Pushrod

The fuel pump was left on until this point to keep the distributor drive in place. To remove it, use a 13-mm wrench to take off the two nuts. Then, pull out the fuel pump pushrod.

39. Remove the Fuel Pump Base

The fuel pump base is made from a Bakelite material and can become stuck in the case. Prying out the fuel pump base usually results in a broken base. Use a large screwdriver to push it out from the bottom. Pry against the distributor drive for leverage.

40. Remove the Distributor Drive

The distributor drive can be pushed out of its bore from the bottom. There is no need to buy a special puller. Once out, keep track of the two thrust washers that are at the base.

41. Clear 3-4 Case Side

The remainder of the little items can be removed from inside the 3-4 side of the case. There are six rubber seals at the base of the main studs, three cam bearing halves, a cam plug, a center main bearing, and four main bearing dowel pins.

42. Clear 1-2 Case Side

A couple items remain in the inside of the 1-2 side of the engine case. These include the center main bearing half and dowel pin and three cam bearing halves. The oil pickup can remain installed.

Parts to Keep and Parts to Replace

While you are in the process of tearing down your engine, there are certain items you are going to replace nearly every time. You don't need to be careful while removing these items but you should hang on to them at least until your engine is complete. Sometimes parts were ordered and you aren't sure if you received the correct replacements. The old parts can be used as a reference.

Once you get the heads removed, there will be a few parts you will replace that you would not in most engines. You will replace the cylinders, pistons, rings, pins, and pin retainers with brand-new units that conveniently come in a kit. The old pistons can be used to practice removing and installing piston rings. It is a skill that only comes with

experience; if you break a ring while practicing on the stuff destined for the scrap bin, you will get the hang of it hassle free.

At this time, all of the pushrod tubes and seals can be trashed.

Once the case is split, all the bearings will be replaced. Bag the old ones and keep them. They are a good indication of the life your engine has lived.

The rest of the internal components will need some cleaning and inspection before determining whether they will be replaced or refurbished. That will be covered in detail in chapter 3.

No matter what condition your engine is in, there are parts you are going to replace during every rebuild. Don't cheap out on the gaskets and seals. Spend the little extra money on German-made products. The last thing you want with a fresh rebuild is a pesky oil leak.

Here is a list of the mandatory parts required to complete a rebuilt long-block:

- Quality German complete gasket kit (Elring)
- German rear main seal (Elring)
- Pushrod tubes
- Main bearings
- Connecting rod bearings
- Cam bearings
- Piston and cylinder kit
- Rebuilt or new cylinder heads
- Rebuilt or new connecting rods

Depending on the condition of the rest of the components, these items may or may not require replacement or reconditioning.

- Flywheel
- Flywheel gland nut
- Camshaft
- Lifters
- Oil pump

PARTS CLEANING AND INSPECTION

The easiest, most efficient time to inspect components is during the cleaning process. By cleaning them and really getting a good eye on most of the parts, you can determine what to do next. We will start with the heads and work our way inward.

Cylinder Heads

Before you even get these near the cleaning tank, a quick visual inspection will tell you if the cylinder heads are worth rebuilding. Broken off exhaust studs are $15 to $20 each to fix. Stripped out spark plug holes are $25 to $30 each to repair. Broken off fins can be expensive to repair. Guides needing replacing will be another $20 to $30 each. All these things will add up.

The cost to clean up, flycut, and cut the valve seats of a rebuildable head can cost anywhere from $75 to $100 each. Add the price of new valves and whatever else the machine shop might find wrong and the price can quickly approach the price of a brand-new replacement head. The price of replacement heads is rather reasonable with a lot of competition in the VW aftermarket.

Rocker Assemblies

Very rarely is there reason for complete replacement of the rockers and rocker shafts. Occasionally, you will run across some really abused adjuster jam nuts, a broken spring clip, or missing parts, but all of these are readily available. (See chapter 5 for the complete rebuild process of these assemblies.)

Pushrods

Pushrods are easy to inspect. Simply roll them on something flat and make sure none of them are bent. You can do this while cleaning them in the parts washer. Replace a bent one with a good used straight one or a new one. Bending them back straight is never recommended. Once bent, they tend to bend again.

New pushrods are available individually or as sets, but road-tested parts work just as well. Rarely are new parts tested for thousands of miles. All of the pushrods are the exact same length and can be replaced in any order. Make sure to blow them out with compressed air. A blow gun with a rubber tip works well.

Replacement heads are readily available at a reasonable cost. These brand-new heads from EMPI (part number 98-1356-B) are designed to fit, run, cool, and last just like the originals. They are available from many online retailers for less than $185 each. (Photo Courtesy EMPI Inc.)

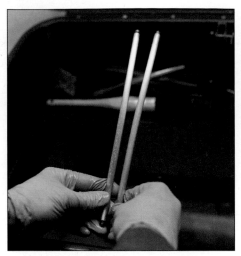

While cleaning the pushrods, we noticed one was bent. The one on the right wobbled when we rolled it in the parts washer. Another used one will have to be found. New pushrods are available (part number 311-109-301A).

Lifters

Sometimes called cam followers, the faces of lifters take a beating because they have metal-to-metal contact with the cam. The faces have a slight radius or dome on them to make them spin as the cam lobe pushes them into their bores. Once the radius wears away, the lifters can be reground, but replacing them is usually more economical.

There is a very simple test to see if the lifters still have a crown on their faces. Clean the face of two lifters, put them together, and hold them up to the light. You should see that only the centers touch and light comes through along the edges. Ours are worn and will be replaced.

This lifter is an example of a galled face. Though the engine ran fine, it was eating itself alive. Metal particles made the engine oil look like metal flake paint. Any time there are sparkles in the oil, it's time to tear it down and find out why. It's not going to fix itself by changing the oil.

Camshaft

The camshafts of VW air-cooled engines are very durable and rarely need replacing. The hardness of the cam is much harder than the lifters, so the lifters seem to take all the abuse.

Use digital calipers to check for cam wear. First, zero out the caliper on the base circle, and then measure the highest point of the lobe. The two center lobes (intake lobes) should be the same height. Do the same thing with the outer lobes (exhaust lobes). Both should be the same. Don't compare an intake height to an exhaust height. These cam lobes have different lift amounts by design.

Connecting Rods

Volkswagen did a great job designing and manufacturing a very stout connecting rod. If the rod bearings are in good shape and the engine didn't have a rod knock which pounded out the rod bearing; the rods are probably usable as is.

A spun or damaged rod bearing along with overheating signs (the big end of the rod is blue) mean you are

To determine if the small ends of the connecting rods are good, push on the piston pins. They should push through with some resistance. Only two of the four did, so we are going to replace the rods.

While cleaning the cam, take special notice of the cam lobes. Make sure none of the surfaces are galled up. If the lifter galled up, so has that lobe on the cam. Cams can be reground but a good used cam may be a more economical choice. New cams are available (part number 113-109-021).

in the market for different rods. Don't replace just one rod; all the rods need to weigh the same to keep the rotating assembly in balance. Rebuilt rods with resized big ends and rebushed small ends are also an option.

Crankshaft

As mentioned in chapter 1, Volkswagen had trouble very early on with crankshafts breaking. The manufacturer doubled down after that to produce one of the strongest cranks ever offered by an OEM. It can be reground undersize without losing any of its original strength.

If you don't feel comfortable determining the size of the crankshaft journals, you are not alone. Accurately reading a micrometer is an acquired skill, and it's of utmost importance that it's done right. We will have the machine shop determine if the crankshaft is useable as is, if it can be polished and still be within spec, or if it needs to be reground undersize.

Flywheel

Transferring the power of the engine to the transmission can take its toll on the flywheel. A slipping clutch can overheat the flywheel, causing cracks that sometimes can't be machined out. A loose flywheel will stretch the four dowel holes to become oblong, rendering the flywheel useless. A sticking starter drive or faulty ignition switch that allows you to "restart" a running engine can destroy the teeth on the outside of the flywheel, making it difficult for the starter to engage. The area where the flywheel seal rides is very important for a leak-free build. A groove there will cause a leak that will only get worse.

Flywheel Gland Nut

Not much can go wrong with the flywheel gland nut. If the threads look good and thread into the end of the crankshaft easily, it's good to go.

Our flywheel has seen better days. Not only is it in need of resurfacing but there is a groove on the backside where the rear main seal rides. This explains why the rear seal had been changed out from the factory black seal to an orange silicone seal. It must have been leaking and a new seal was an attempt to fix it.

Worn-out and defective flywheels can cause a host of problems from leaking crankshaft seals to broken starter drive gears to clutch chatter and failure. A new replacement like this one from EMPI (part number 98-1273-B) can remedy all those things for about $65. (Photo Courtesy EMPI Inc.)

Thoroughly cleaning the crankshaft at a well-lit, well-ventilated parts cleaner can reveal some sins of the previous owner. Ours looks rather nice. There are no signs of overheating, which show up as blue areas usually at the rod journals. There are also no ridges on the journals that could be caught with a fingernail.

Check the oil pump with feeler gauges to measure the end gap of the gears. The gears should be flush to 0.004-inch gap without the gasket. It's important you use the correct gasket and not a homemade one. The thickness is critical.

The only problem area is the pilot bearing that is found inside the head of the nut. This small roller bearing supports the input shaft of the transmission. A small felt seal is held in with a small tin shield. If the shield or the felt seal are damaged or missing, a new gland nut is in order.

Oil Pump

If your engine was lucky enough to have regular oil changes, your oil pump shouldn't need replacement. The oil strainer at the oil pickup is the only defense from items other than oil entering the oil pump housing. If the housing has deep grooves along the perimeter, where the gears ride or the gears themselves are galled up, a new oil pump is a good idea.

To test an oil cooler for leaks, make a tester from an oil cooler adapter. These adapters are usually used to install an external oil cooler in an off-road application. Plug one hole and add an air fitting to the other.

Oil Cooler

The oil cooler is often overlooked. If it doesn't have a leak and your engine didn't experience a major catastrophe, then it's safe to reuse. If your engine spun a bearing, seized up, or had metal particles in the oil, then a new oil cooler is mandatory. No amount of cleaning or flushing will guarantee all the tiny specks of metal are out of the cooler.

Early steel oil coolers are expensive but still available. Late-model doghouse oil coolers are inexpensive. Unknown status oil coolers can be tested by using a special, easy-to-make tester.

Engine Case

Just like the foundation of a house, the engine case will determine the outcome of a build. It will be foolish to compromise on any aspect of this component. The first thing to check is if it has overheated. Air-cooled engines don't "boil over" like water-cooled engines do. Volkswagen never had any gauges or warning lights to warn if things were going wrong. They just keep running and usually do damage along the way.

An overheated engine will expand to the point that it loses torque on the main studs and warps.

Mount the tester to the bottom of the oil cooler. Open up the holes in the adapter to 5/16-inch for the late-model coolers. Use the appropriate seals for the cooler from the gasket kit.

Attach a regulator and submerge the cooler in a bucket of water that has a couple drops of liquid dish soap in it. Gradually add air to the tester while watching for bubbles. It should only take 40 to 50 psi to confirm whether the cooler passes.

If your oil cooler has a leak, new coolers are readily available. This unit from EMPI (part number 98-1161) is an exact replacement and only costs around $40. Early in-shroud coolers are about $100 new. (Photo Courtesy EMPI Inc.)

To prep an engine case for initial inspection, remove all cam bearings, center main bearings, and six main stud seals. Next, scrape both halves with a razor blade to remove sealer that's usually baked on.

The center main bearing area will spread apart, and since the torque is lost on those studs, when it cools back down it won't stay together. This is commonly known as a "split web." An engine can live like this for quite a while without giving any signs of the damage other than decreased oil pressure when up to operating temperature and more oil leaks than usual.

To check for this problem, the first thing to do is clean the mating surfaces of the two case halves with a razor blade and then a flat file. Next, torque the case together. Only the six larger main studs need to be torqued. With a small flashlight, shine light through the crankshaft hole on one end while looking for light to shine between the two halves at the center main bearing area.

Another way to check is with a feeler gauge. Use the smallest one you have and try to slip it between the two halves at the center main area. If you see light or a feeler gauge

Reinstall the six main stud nuts and washers. Use a squirt of oil on the threads and on the washers before spinning on the nuts. It's not necessary to install any other hardware at this time.

fits between the two halves, the case is unusable and unfixable. The only other option is a different case, either new or used.

Next, use a flat file on both halves to make sure nothing is raised up to prevent the halves from matching up perfectly. If the engine was apart before, there may be pry marks from a screwdriver wielded by an overzealous mechanic trying to get them apart.

With a 3/8-inch drive torque wrench and 17-mm socket, torque the six nuts to 25 ft-lbs. Start with the center two nuts and work outward in a cross pattern. Check for a split web at the center main bearing area with a flashlight or a feeler gauge. Any gap there will render this case scrap, and you will have to find another case.

The area circled is prone to cracking and is hard to diagnose. If the back of the case and the back of the flywheel are covered in oil and if it looks like the rear main seal has been replaced numerous times in recent history, inspect this area very closely. Spray the area with brake cleaner and blow it off with a blow gun, which will usually get the crack to weep oil. A crack here is definitely unfixable.

Another serious (but less common) problem is a cracked case on the pressure side. The 3-4 side of the case carries the majority of the oil pressure, which means it will have the majority of the oil pressure–related cracks. The two most common areas are: the oil galley that feeds the last cam bearing behind the flywheel and the oil pressure sender hole at the top of the case near the distributor. Less common areas for cracks are between the head stud holes and the barrel hole or internal oil passages. Take a good, hard look at all these areas while cleaning the case. Good lighting is key. Grab the flashlight!

Lifter Bores

The lifter bores are especially important for a quiet-running, long-lasting engine. Worn lifter bores will bleed off too much oil pressure in a hot engine and cause premature bearing failure. Less oil gets up the pushrods to the rockers, causing a very "ticky" engine especially

Check the lifter bores in the engine case for excessive wear. With a new lifter, pull the lifter out approximately 3/8-inch and push it back and forth. It should only have a slight amount of play. The recommended high limit is 0.0047-inch movement, or a maximum bore diameter of 0.7499 inch.

at idle with a hot engine. It is not uncommon for all the lifter bores to be good with the exception of one or two bad ones. If any are worn beyond the wear limit, the bores

The oil pressure sender hole is an often mistreated area. The sender has tapered threads, so it doesn't take a lot of force to seal the fitting. Some people mistake this for straight threads and keep tightening this fitting until either the case cracks or the hole strips out. Either way, welding this area up and reestablishing that hole is the only way to fix it correctly.

can be sleeved and remachined, but this is expensive and finding a better case to start with might be a more economical way to go.

MACHINE SHOP

Professional machine shops have expensive equipment and do excellent work, but it comes at a price. That expensive equipment needs to pay for itself. Precise valve seat grinding equipment and crankshaft, camshaft, and flywheel grinding equipment are great, but sometimes the price of a new or remanufactured part will be more economical than having parts reworked.

Choosing a Machine Shop

As with any engine rebuild, finding and dealing with a machine shop will be necessary to achieve your ultimate goal of a freshly rebuilt engine. Air-cooled VW engines require specialized equipment and procedures to be machined. If your local automotive machine shop says it doesn't have the equipment necessary to machine your out-of-tolerance components, ask around!

Ask a local VW club or join some online VW forums and ask other enthusiasts who they use and who they recommend. Most areas of the country have a VW-specific shop that specializes in air-cooled engine machine work.

If you live in the middle of nowhere and can't find a local shop to do the work, ship your parts to a shop and pay to have them shipped back to you after the work is done. Shipping will not be cheap. This cost should be taken into consideration when budgeting your build. If your case or heads are going to take a lot of work to get back to running order, you may consider buying new parts instead of shipping your parts across the country twice. A new set of heads may be only a couple dollars more than having your heads rebuilt and shipped back and forth.

No matter where you are taking or sending your engine components for machine work, make sure you get a solid quote on services and return shipping if necessary. Also try to nail them down on a time line. If they say three to four weeks, make sure you hold them to it. The last thing you want to tell them is that you are in no rush. You might never get your parts back. Be realistic.

Engine Case Machining Procedures

The engine case is divided in two halves directly down the middle vertically. As it is installed in the vehicle, the right half contains the number-1 and number-2

Notable Machine Shops

The most notable machine shop dedicated to VW air-cooled engines in the nation is Rimco Machine. Established in 1962, Rimco has an impeccable reputation for quality and its employee base has an average of 22 years with the company. In 2017, another staple in the VW industry for 47 years, FAT Performance, bought Rimco Machine and continues to machine and provide service for customers worldwide.

Another notable machine shop is Car Craft in Riverside, California. In business for more than 30 years, it is the owner of the original EMPI machine shop. It specializes in turbocharging, tuning, engine repair, machine work, fabrication, and much more.

All the machine work performed for the engine in this book was provided by Underdog Racing in New Berlin, Wisconsin. It is a one-man shop that specializes in VW Type 1 engines and transmissions. ■

cylinders and the left half contains the number-3 and number-4 cylinders. The odd numbered cylinders (number-1 and number-3) are closest to the flywheel.

The right side is sometimes referred to as the suction side due to the oil pickup that feeds the oil pump being attached to that side. The left side is referred to as the pressure side due to the fact that most of the oil pressure is contained in that side. The oil cooler and oil pressure relief valves are fed from that side.

For the sake of clarity, the terms "1-2 side" will refer to the right side or suction side of the case. The term "3-4 side" will refer to the left or pressure side of the case. The terms left or right get blurred when spinning an engine on an engine stand. No matter how the engine is positioned in the stand, the 1-2 side and the 3-4 side will always be clear.

Case Line Boring

Unless you are building an engine with a new case, it's more than likely your case is going to need the main bearing journal areas remachined oversize. A special boring tool called a line bore bar is necessary to machine these areas accurately.

The boring bar is self-contained and operated by an air-powered drill. It is supported at both ends of the bar. The drive head is supported on the rear main seal area and the other end is supported by a collar tapped into the case where the crank pulley clears the case. Four preset cutters machine all four bearing journals at once. Standard case size at the three large main bearings is 2.559 inches (65.00 mm). The cutters start at 0.020 inch (0.50 mm) oversize and go up in size by 0.020-inch (0.50-mm) increments.

Though it is rare, an engine case can be line bored up to 10 times, going to 0.200 inch (5.00 mm) over the outer diameter (OD) bearings. Other factors usually render the engine case useless beforehand, such as a cracked or split center main bearing area, but 0.080- to 0.100-inch oversize OD main bearings are becoming more common.

A quick check with digital calipers will give you a good idea of what size your case was when it was put together last (or originally). If you don't possess the skills to determine what size it is now and what it needs to be cut to, don't worry about it. The machine shop will determine what size it needs to be cut to. The bearing surfaces need to clean up 100 percent on all four journals.

The line bore bar can also perform a thrust cut. The thrust bearing (nearest the flywheel) supports the crankshaft on its rotating axis and keeps the rotating assembly from moving in and out of the case every time you push in the clutch. With use, this bearing becomes loose in the case and pounds out the edges that support this bearing. The good news is, bearings can be bought with more material built into those surfaces.

Along with bearings being available oversize every 0.020 inch (0.50 mm), starting with the 0.040-inch (1.00-mm) oversize bearings, you can get bearing sets with either 0.040 or 0.080 inch (1.00 or 2.00 mm) added to the inner thrust area. That means the case will need to be machined either 0.040 or 0.080 inch (1.00 or 2.00 mm) on the thrust area to fit the oversize thrust of the bearing. This dimension in the case is 0.785 inch (19.94 mm) and has a wear limit of 0.001/0.002. Anything more and the thrust area of the case will need to be machined. Again, if you are unsure of what is acceptable, the machine shop should advise you if this needs to be done or not.

1. Case Line Boring Bar

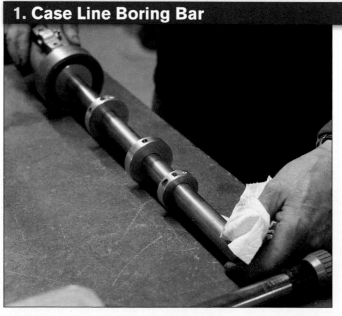

This VW case line bore bar was made by Port-A-Line. It has four preset cutters that cut all the main bearing journals at once. Different oversize bearings require different size cutters. The bar is driven by an air-powered drill that spins the cutters and also drives a pump that feeds the bar through the case.

2. Measure the Bores

Measure the main bearing bore to determine what size it is and what size cutters will be necessary to get the bores to clean up. This measurement doesn't need to be precise, just close enough to get an idea of how badly the bores are worn.

3. Install the Boring Bar

Once the six main studs are torqued to 25 ft-lbs and the two 8-mm nose studs are torqued to 14 ft-lbs, the boring bar can be gently installed into the case. The case was already cleaned up on the sealing surfaces from the inspection process.

4. Bore the Case

The case is being bored 0.020-inch (0.50-mm) over standard size. The thrust area was checked and had only minimum wear. The 0.020-inch over bearings only come with standard thrust. If the thrust was undersize, we would be forced to cut the main bores 0.040 inch (1.0-mm) over because that's where thrust cut bearings begin.

5. Remeasure the Bore

By setting a micrometer with the actual bearings being used, we will get a true reading for setting the bore gauge. Set the bore gauge to zero with the micrometer locked at the bearing size.

Check each main bearing bore with the bore gauge. We are looking for bearing crush or interference fit from 0.0015 to 0.002. Too much crush (say 0.0025 to 0.003 inch) will pinch down the inner diameter of the bearing and may cause the engine to seize up prematurely. Not enough crush (0.001 inch or less) will limit the life of the engine.

Installing Case Savers

Installing threaded inserts, or case savers (as they are called), is common practice to solve a frequent problem with all Type 1 engines with 10-mm head studs. Over time, the head studs tend to pull out of the soft magnesium case and cause a compression leak. With one stud loose, the others soon follow, and a complete engine teardown would be needed to fix the initial cause.

Case savers do exactly what the name implies. Installing these threaded inserts at the time of the rebuild could prevent a head stud from pulling out and ruining your fresh engine. These inserts come in two sizes for 10-mm head studs. In 1973, beginning with the AH serial number cases, VW changed the head stud size from 10 mm to 8 mm. These smaller studs were installed into factory case savers and this (along with the lower head torque required) solved the head stud pulling problem from then on. The later cases actually held their head torque better by allowing more stud stretch and fighting the constant heating and cooling effects of the engine.

1. Case Savers

Two different size case-saver inserts are shown: 1/2-inch-20 OD (left) and 14.0-mm-2.0 OD (right). Both use 10-mm head studs. Case saver is an industry term for the threaded insert that is used to install the head studs into the magnesium case.

2. Drill Out the Stud Holes

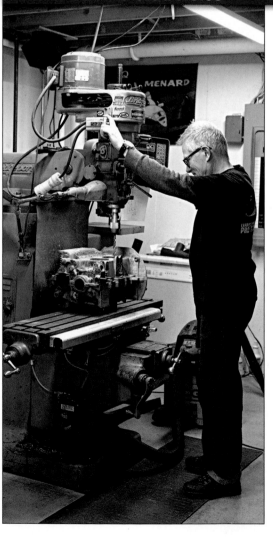

The first step in installing case savers is drilling out the existing head stud holes. A Bridgeport mill makes sure they are all square and true, but a good drill press can achieve the same results. We are using 1/2-inch-20 case savers that require a 29/64-inch drill. The drill will follow the existing holes, and no cutting fluid is required.

3. Tap the Holes

Next, we power tap all the holes with the 1/2-inch-20 tap. The thru holes we can power tap all the way through. The blind holes we stop short and finish tap by hand. The last thing you need is a broken tap. Always use a tapping fluid specific to nonferrous materials such as aluminum and magnesium.

4. Install the Case Savers

Finally, we install the case savers. Each insert is screwed onto a 10-mm bolt with a jam nut. Red Loctite is put on the threads and then threaded into the freshly tapped holes. An impact driver makes this chore much easier. Make sure the insert is flush or lower than the surface the barrel seats on.

Checking Stud Torque on Rear Cam Journal Studs

1. Oil the Threads

The two 8-mm studs that straddle the engine mount stud on the 1-2 side of the engine have a strange tendency to pull out. Any time before final assembly, make sure they are up to the task. Apply a couple drops of oil on the threads before using the studs.

2. Apply Torque

The only two studs we are worried about are the ones shown with the nuts on. Torque those two to 15 ft-lbs, which is one more pound than the required torque. If the studs hold the torque, all is good. If the stud pulls out before the required torque, then the tapped hole will have to be repaired with a 8-mm-1.25 Heli-Coil.

Converting Case to Full-Flow Oiling

Converting a case to full-flow oiling isn't mandatory, but if you decide to add an external oil filter or auxiliary oil cooler or both in the future, now is the time to lay down the initial machine work. Full-flow is the industry term for rerouting the oil from the pressure side of the oil pump, out a port in the oil pump cover, and back into the case through a port drilled and tapped just above the oil cooler relief passage.

A standard pump can be used and the inlet port plugged until you do the rest of the conversion at a later date. Many cases have been ruined trying to drill and tap this port while the engine is still together. It's even more difficult with the engine still installed in the car. If you think this is something you will do in the future, now is the time.

1. Drill the Oil Passage

With the case halves bolted together, clamp the case to the table of the Bridgeport mill. Using a 9/16-inch end mill, drill out the plug of the oil passage just above the oil cooler bypass. Send the end mill deep enough to tap the hole all the way but be careful not to get into the area that the piston for the bypass seats on.

2. Tap the Holes

Tap the hole with a 3/8-inch-18 NPT tap. Use a center in the spindle to keep the tap going in straight, and use cutting fluid specific for nonferrous materials. Never use a dull tap; it will put too much stress on the material and cause the case to split.

Tap the hole just deep enough to get three quarters of your fitting to thread in. Overtightening this fitting will crack the case and ruin it. Never use Teflon tape on these fittings; instead, use green Loctite and only slightly tighten the fitting. The thread locker will seal the threads without overtightening the fitting. Some clearancing to the boss adjacent may be necessary to get the fitting to clear.

Head Rebuilding Procedures

Head rebuilding has many aspects. First and foremost are the valves and their components. Always replace the exhaust valves. Period. You have no idea how much abuse they have endured and they are inexpensive to replace, so just do it.

The intake valves can be reused if the shaft is not worn or pitted, but they will always need to be resurfaced. Intake valves are also inexpensive, but for the budget minded it's alright to reuse them once resurfaced.

Next are the valve guides. These are sometimes a pain to replace, but they should never be overlooked. The factory has a wear limit of 0.0031 inch (0.08 mm), but that is excessive. A wear tolerance of 0.0010 to 0.0015 inch (0.025 to 0.030 mm) is more

commonly accepted. Again, if unsure, ask your machine shop what it recommends. Most machine shops reuse the stock valve springs, retainers, and valve keepers. All of these items were well engineered from the factory and made of quality materials.

Now we get into the head casting itself. An air-cooled VW head is the most abused component of the entire engine. Operating at 300 to 350°F takes its toll on the cast aluminum, and many heads develop cracks. The combustion chamber is especially susceptible to cracks between the valve seats or from the valve seat to the spark plug hole. Cracks are signs that a head has come to the end of its usefulness and should be replaced.

The first thing any VW machine shop will do, even before putting the heads in the parts washer, is sandblast the combustion chambers. This will determine if the head is even rebuildable. Any cracks between the valve seats or between the seats and the spark hole will render the head unsavable.

Any rebuildable head will need to be flycut during the rebuilding process. During the operation of the engine, the head stud torque will be tested with every combustion cycle. During the many heating and cooling cycles during an engine's lifetime, the head stud torque may lessen and the head will begin to pound into the sealing surface of the top of the barrel.

To reestablish a flat surface for the barrel to seat on, the head will need to be remachined. A minimum amount of material should be removed to avoid changing the compression ratio. A stock head should have a 51-cc combustion chamber. (See chapter 6 for the procedure to calculate compression ratio.)

Head Disassembly

Special Tool

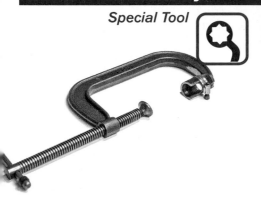

1. Apply the C-Clamp

Position the clamp and pipe with the notch up. The hose clamp on the pipe hangs over the edge slightly and helps it stay on the clamp. Compress the spring with the clamp.

A special tool can be purchased to remove the springs from a VW head, but most are poorly made and require you to do everything one-handed because the other hand is tied up compressing the spring. An 8-inch C-clamp and a small piece of pipe with a notch in it work just as well and allow you to use both hands.

2. Remove the Keepers

With a small magnet, remove the two keepers and place them in a little cup for safekeeping. Put the springs and retainers in another basket. It's not necessary to keep track of which valve these items came from; they are all the same and can return in any position. Leave the valve in the guide for now.

It's easy to pull the keepers out with a tiny magnet. Sometimes you may need to pick them off the valve with a small screwdriver. Once the spring is compressed, you can use both hands.

3. Remove the Valves

Once all the keepers, retainers, and springs are removed, the valves should slide out. If they hang up on the guide, do not force them through. There is a burr on the valve from the keepers riding in the three grooves. You need to file off that burr while spinning the valve with your fingertips. It'll slide right out.

Initial Preparations

1. Chase the Threads

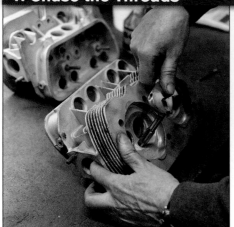

Unless you are using new heads, chasing out all the threads is a good idea. Run a 14-mm-1.25 tap through the spark plug holes and a 6-mm-1.0 tap into the holes that hold on the cylinder tin. The last thing you want is to find out during assembly that you have an issue with either of these.

2. Chase the Studs

While chasing the spark plug holes and cylinder tin holes with a tap, now is a good time to chase all the exhaust studs with a 8-mm-1.25 die. If the threads are too damaged or rotted away, apply heat (a lot of heat) and try to unscrew the stud and replace it.

Exhaust Stud Repair

Special Tool

The 8-mm-1.25 thread repair kit is the most popular thread size on any VW by far. The intake and exhaust studs are this size along with all of the perimeter case studs, carburetor, generator stand, and much more. This kit retails for $27 and is a must-have item.

1. Stud Repair Attempt

Here we have a poorly repaired exhaust stud hole. The stud must have snapped off and an attempt to fix it was to drill it smaller and retap it for a smaller thread size. We will repair it correctly with a thread repair kit, which is sometimes called by its trade name: Heli-Coil.

2. Drill the Hole

Once clamped to an angle plate in the Bridgeport mill, we will drill out the hole to the correct size for our kit. In this case it's a 21/64-inch drill.

3. Tap the Hole

Start the special tap by putting it in the chuck and power tapping it the first couple threads to be sure it starts straight. Use cutting fluid specific for nonferrous materials. Always hand tap the last few threads of a blind hole. Broken taps are very difficult to remove.

4. Install the Insert

Install the thread insert with the special tool included in the kit. Install a new stud with red Loctite and the process is complete.

Cylinder Tin Mount Repair

1. Drill the Hole

The cylinder cooling tin is held to the head by two 6-mm-1.0 screws. Too often these screws get snapped off during the removal process. We will repair them with a thread repair kit similar to the exhaust stud repair but in a different size. First, we will center drill the broken screw and drill it out with a 1/4-inch drill.

2. Tap the Hole

Next, we tap the hole with the special tap from the kit. Start the tap by putting it in the chuck and turning the chuck by hand. Once you get a couple threads in, complete the tapping process by hand. Use cutting fluid specific for nonferrous materials.

3. Install the Thread Insert

Install the thread insert in the installer included in the kit. The coil insert will screw out of the installer and into the freshly tapped hole. Make sure the insert is flush or slightly below.

Flycutting the Head

1. Deburr the Valve Cover Surface

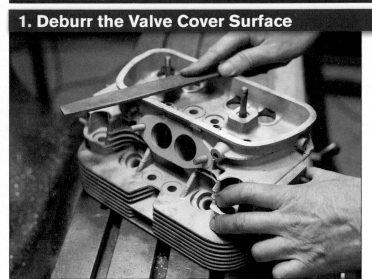

Before clamping the head to the Bridgeport mill, it is always a good idea to make sure the valve cover surface is free of burrs and dents. People tend to get a little overzealous when removing the valve covers. The rocker studs should clear the bottom of the T-slot in the table, but double-check to be sure.

2. Set Digital Readout

Swing in one bore with a dial indicator mounted in the spindle and zero out the digital readout. Go to the other bore and swing that one in as well. It should be roughly 4.410 inches away from the first bore. Write down that location.

3. Cut the Head

Lock the table and head and raise the table to cut just the surface the barrel seats on. Once you achieve a completely cleaned up surface, zero out the dial that raised the table. Repeat the same procedure in the other bore. Make sure not to unlock the head between the two bores. But, be sure to raise the table up to the exact same height in both bores. This will ensure both surfaces are at the same depth.

Replacing Valve Guides

1. Check for Wear

Check the valve guides for excessive wear. Use a new valve to pull it out of the guide roughly 1/2 inch. Try to rock it back and forth. The most it should move is 0.0010 to 0.0015 inch. Anything more than that and valves guides will need to be replaced.

2. Tap the Guides

The exhaust valve guides are too worn out and need to be replaced in this head. The first step is to tap the rocker box end of the guide with a 3/8-inch-18 tap. Tap as deep as you can.

3. Insert the Bolt

Thread a 3/8-inch bolt into the guide and make sure it is bottomed out. Tap and thread all the guides that need to be replaced at this time.

4. Punch Out the Guide

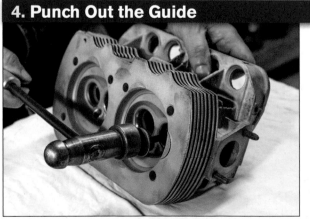

Use an old exhaust valve that was removed from the head. Drive out the guide with a good-size hammer, such as this 32-ounce dead blow ball-peen hammer. It might take some force. If you didn't tap the guide deep enough, you may just push the bolt out without the guide.

Here the guide finally comes out. It measured 0.475 inch, which is the factory replacement size, so standard ones should go in just as hard as these came out. They also come 0.002, 0.003, 0.005, and 0.008 oversize to fit bagged out holes in the head.

5. Guide Installer

Our handy guide installer consists of a 5/16-inch socket head cap screw that is 3 inches long and has a threaded bushing to match. The bushing is 1 inch long and 0.473 inch in diameter to slip in the guide hole and get the guide going dead straight.

6. Heat Up the Head

Heat the head up in an oven to 400°F. This will allow the guide to press Into the head easier. Cool the guide and installer assembly in the refrigerator or outside (if it's cool out). Apply assembly oil to lube the guide.

7. Install the Guides

Quickly retrieve your chilly guides and warmed up head and drive the guides into the head with an air impact hammer. Use heat resistant gloves and eye protection. The proper safety equipment is a must.

Cutting Valve Seats

There are varying processes to resurface valve seats. Some use very sophisticated and expensive equipment to achieve very precise tolerances. Most professional machine shops use such equipment because it performs the job quickly and easily. However, a premium is charged for this.

There is a more "hands on" way to resurface the valve seats. The system we are using is produced by Neway Manufacturing in Corunna, Michigan. It has been manufacturing valve seat cutting equipment for more than 50 years. Many smaller shops use this system because it is versatile. Though it does take longer to cut the seats with three individual cutters rather than one very specific cutter that cuts all three surfaces at once, the results are the same.

A good amount of skill is involved to achieve the proper valve sealing properties, but the amount of equipment is minimal. All that is needed is one pilot (PEM150-8mm), two cutting heads (CU230 and CU206), and an EZ turn wrench (NTW-EZ) to do a complete three-angle valve job. To purchase this equipment new isn't too outrageous; the total for all four items is less than $250.

1. Install the Pilot

First, install the pilot (part number PEM150-8mm retails for $43.25). This expanding-type pilot locks into the guide and is self-centering in slightly worn valve guides.

---

2. Make the 45-Degree Cut–Exhaust

Next, establish the 45-degree sealing cut on the exhaust seat. This may take some effort to get to 100-percent clean. The cut is made with the 45-degree side of the Neway cutter (part number CU230 retails for $114.50).

Make sure that the 45-degree cut is cleaning up around 100 percent of the seat. Sometimes it's hard to see, and marking the seat with a permanent marker can help determine if you are deep enough.

3. Make the 45-Degree Cut–Intake

Cut the 45-degree sealing surface of the intake seat. By cutting all of the 45-degree surfaces on each valve seat first, you can determine whether any of them will need to go deeper to even out the depths.

4. Make the 30-Degree Cut

Then, make the 30-degree top cut on each seat to establish the outer diameter of the 45-degree surface. This is done with the 30-degree side of the same Neway cutter (part number CU230). The transition between the two angles needs to be slightly smaller than the diameter of the valve.

5. Make the Final 75-Degree Cut

Finally, go back in with the 75-degree cutter (part number CU206 retails for $70.25) to establish the finish width of the 45-degree surface. It should be between 0.055 and 0.098 inch wide.

Spark Plug Hole Repair

Special Tool

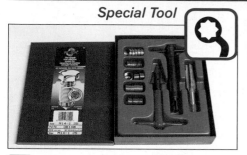

This spark plug hole repair kit has everything necessary to repair 14-mm-1.25 threads. All of the engines covered in this book have 1/2-inch reach spark plugs with this thread. This kit retails for around $50.

1. Inspect the Head

The spark plug hole threads in this head are barely there. As long as there are some remnants of threads, they can be repaired using a Heli-Coil type repair kit.

2. Install the Tap

Install the special spark plug Heli-Coil tap into the damaged hole. The first part of the tap is 14-mm-1.25, so it should screw in rather easily while chasing the messed up threads. The next little part of the tap cuts the hole larger for the coil insert.

3. Check the Spark Plug Angle

Once the pilot part of the special tap is in all the way, screw a spark plug into the good side (hopefully there is a good side). Use the spark plug as a gauge to make sure you are threading the tap into the hole at the exact same angle. Use cutting fluid specific for nonferrous materials.

4. Install the Coil Insert

Slide the coil insert onto the special tool and thread them into the freshly tapped hole. Do not use any Loctite or thread sealer. If done correctly, the coil won't unscrew with the spark plug.

5. Snap Off the Driving Tab

Place the very end of the insertion tool on the end of the coil insert you threaded into the hole. Hit it sharply with a hammer to snap off the driving tab.

6. Final Inspection

Your freshly repaired spark plug hole is as good as new; some say even better. Be sure to use anti-seize compound on all the spark plugs when installing them. This will help prevent any spark plug hole thread issues in the future.

Head Reassembly

1. Gather All Parts

Once all the machine work is done (the head is flycut, guides are checked or replaced, the seats are cut, any problems fixed, and everything is cleaned up), we can reassemble the head. Always use new exhaust valves. We are using new intake valves as well. Unless there is an obvious problem, such as a broken spring or retainer, all the valvetrain parts can be reused.

2. Lubricate the Valve Stems

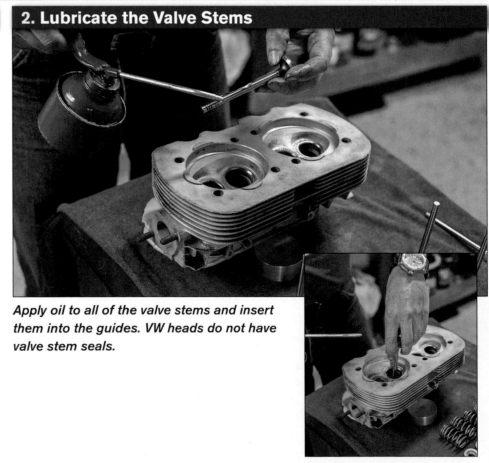

Apply oil to all of the valve stems and insert them into the guides. VW heads do not have valve stem seals.

3. Install the Valves & Hang the Springs

Spin the head around and hang all of the springs on the valve stems. The compound rate springs have a top and bottom. The tighter wound end goes toward the base of the spring pocket. The looser wound end is at the retainer.

4. Compress the Springs

With a retainer on top of the spring, compress it until you can see the three grooves that hold the keepers. The end of the spring wire should face up when installed, this way you can see if it snapped off. You should check that every time you adjust the valves in the car.

5. Install the Keepers

Apply some oil or grease to the keepers and place them in the grooves. Putting the keepers on the end of a magnet works well to get them seated correctly.

6. Check Keeper Installation

Loosen up the C-clamp and watch to make sure the keepers don't fall off. Repeat this on the last three valves to complete the head.

7. Final Inspection

Here we have all the valves installed and the retainers and the keepers are in place. Double-check to be sure that all of the keepers are seated correctly in the retainers. It's possible to put one in upside down. Also, make sure the keepers touch each other at the tips. Some aftermarket valves had the grooves too shallow and the keepers clamped on the valve before the sides touched, keeping the valve from spinning once installed.

Post–Machine Shop Procedures

Once you get all your components back from the machine shop, it's wise to run everything through the parts washer again and blow everything off with compressed air.

Try to do this close to the time you will be reassembling the engine so the chance of more contamination is at a minimum.

The case is especially important to clean out thoroughly. Chips from the line bore process get in everywhere, and the full-flow pro-

cess puts chips directly into the oil passages. A reground crank could have grinding sludge still in the oil passages. Most machine shops won't guarantee everything is cleaned out completely. The final cleaning and inspection is entirely on you, the engine builder.

Post–Machine Shop Procedures

After all of the machine shop processes are completed, it's time to do a thorough cleaning. Even if the machine shop says it ran the parts through the industrial parts washer, you'd be wise to give everything another cleaning.

After another trip to the parts washer, blow everything out once more. If you are using a water-based cleaning solution, rinse everything off with hot water and blow everything dry. These solutions tend to leave a residue that doesn't evaporate like solvent-based cleaners.

COMPONENT PREPARATION

Rebuilding an engine is not an easy task. It takes planning, and planning takes time. If you want to keep your project on track, it always helps to use your time wisely. While some of your parts may be at the machine shop or you are waiting for new parts to arrive, now is a good time to focus on what you can work on.

You are going to have your hands on every component of the build. A decision has to be made on what to do with each item. Sometimes the decisions are easy and sometimes some thought has to be put into what to do.

One question that comes up is "Should I even bother rebuilding the engine I have?" This is often the case when the engine isn't exactly what you want. Say the engine you have to rebuild isn't the correct one for your vehicle. It's older and you'd like the right year or possibly a newer model. Now it isn't feasible to convert a 40-hp 1,200 into a 1,600 dual-port, but it is possible to covert a single-port into a dual-port. In fact, no extra machine work is necessary to convert a 1,500 single-port into a 1,600 dual-port. It's just a matter of gathering the right parts.

You already need to buy a new set of pistons and barrels, and truthfully, the 85.5-mm (1,600-cc) kit will be cheaper and easier to find than a 83-mm (1,500-cc) kit. Your single-port heads might be beyond repair, so new heads are in order anyway; you may as well get new dual-port heads that are roughly the same price as single-port heads.

You will need dual-port cylinder tin and the dual-port intake manifold. Your 30PICT carburetor can be adapted to the manifold or you can upgrade to the 34PICT carburetor. There are a couple more minor factors, but these can all be taken into consideration when deciding what route to take.

Single-Port to Dual-Port Conversion

To install the dual-port intake manifold center section on a single-port case, you need to swap out one of the perimeter case studs for the one shown in line with the alternator stand studs. It's roughly 20 mm longer than the early one.

Dual-port heads require three different head stud lengths. The single-port had two lengths (center-right). Swap out the upper center studs (center) with the two shorter ones (left) to convert to dual-port heads. Two 203-mm studs get changed out for two 187-mm studs on each side of the engine.

How to CC a Cylinder Head *Continued*

1. Level the Head

Set the "ready to install" head on a
nice level bench (our Bridgeport mill
is the most level thing in the shop).
Install a spark plug in one chamber
and set the head on a couple blocks.

2. Seal the Disc

Apply a little white grease or petroleum jelly around the edge of the
disc. This will create a good seal
between the disc and the mating
surface of the cylinder in the head.

3. Add the Solvent

4. Add More Solvent

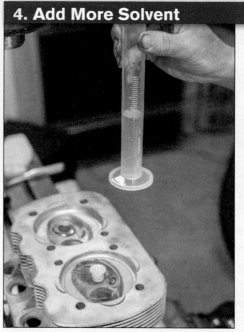

We put 40 cc into the cylinder during the first fill. Now with exactly 30 cc in
the cylinder, we can finish filling the combustion chamber. We will add solvent
until we have just enough in to get rid of the air bubble but not overfill.

5. Measure the Solvent

The chamber is completely filled with fluid, and now we can do a little math.
The first fill was 40 cc. We had 18 cc left of the 30 cc fill once we got the air
bubble out of the chamber ($30 - 18 = 12$ cc). Add 12 cc to 40 cc, and we
have 52 cc total.

With the disc sealed to the head, pour parts washer solvent from a graduated
cylinder into the chamber through a tiny funnel. The graduated cylinder only
holds 50 cc of fluid. A stock head should be around 52 to 53 cc, so we will need
more to top it off.

Rebuilding the Rocker Assemblies

1. Gather the Components

Here are all of the components for the rocker arm assemblies. Disassembly and cleaning is pretty straightforward. All the parts are interchangeable. We washed everything in the parts washer and bead blasted the rocker arms and blocks.

2. Polish the Rocker Shafts

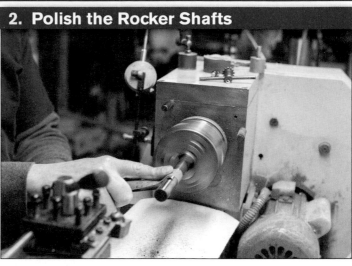

If you have access to a lathe, polish the rocker shafts to remove all of the scale and imperfections. After a trip to the parts washer, some 400-grit emery cloth was used to make them shiny again.

3. Chase the Threads

You will need the following parts for one assembly: one shaft, two blocks, eight flat washers (four of each), rockers, adjusters, jam nuts, spring washers, and clips. Here, we are chasing the threads of each rocker with a 8-mm-1.00 tap in a cordless drill on low speed.

4. Create Assembly Stand

A large tapered punch is clamped in the vise and one rocker shaft and block are slid down on it. This will be our assembly stand to build one rocker assembly. Note that the slot in the block is away from you.

5. Add Adjuster Screws

Apply some oil on the adjuster screws and thread them into each rocker arm. They should thread into the rocker by hand.

6. Add Jam Nuts

Add a jam nut to each adjuster. Replace any nuts that look like they've had a wrench slip off them too many times.

7. Oil the Rockers

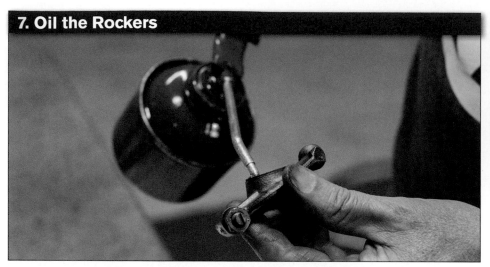

Oil each rocker before sliding it onto the shaft. It will go on the shaft with the adjuster away from you. Both the slot in the block and the adjusters are up when installed in the engine.

8. Add Washers

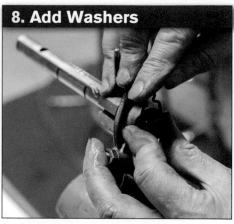

Make a sandwich of washers (flat washer then wave washer then another flat washer) and add it next. Compress these against the rocker and push on the clip.

9. Final Assembly

Repeat this process on the other side of the block. Slide on an oiled rocker, add a sandwich of washers, and then push on the clip.

The other end of the shaft can get everything stacked right on. Start with a clip toward the middle. Then the flat washer, wave washer, flat washer sandwich.

Next add a rocker, then a block, the last rocker, the final sandwich of washers, and finally the clip on the end.

Long-Block Cooling Tin

Air-cooled engines in general have a lot of tinwork, either to keep areas separated or to direct the flow of air. Keeping air moving to places it needs to be and keeping it away from places it shouldn't be are very important for the longevity of any air-cooled VW engine.

The sheet metal needs for your rebuild will vary greatly, depending on many factors. The end goal needs to be known at some point in time. That's where the sheet metal differences are going to make it work efficiently.

An air-cooled VW long-block, valve cover to valve cover, flywheel to crank pulley, can fit into a vast array of vehicles from showroom stock Beetles to dune buggies to sand rails to log-splitters to water pumps or even Zambonis. You name it, and someone has powered it with an air-cooled VW

engine. The base of all these machines remains the same; the cooling system must match the needs of the machine.

Long-Block Engine Tin Preparation

Maybe you haven't made up your mind on all the cooling tin options. There are a lot of items that can be prepped to keep the build on track. These items are necessary to put the long-block together and can be done at any time before reassembly.

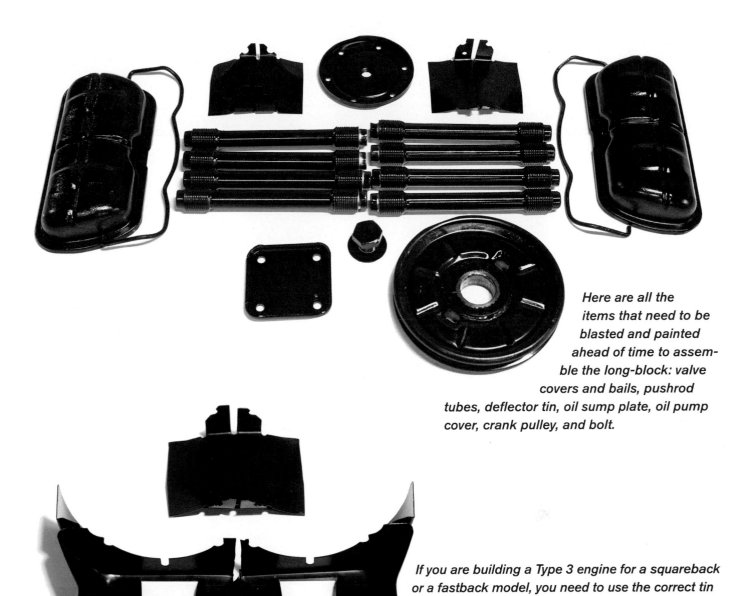

Here are all the items that need to be blasted and painted ahead of time to assemble the long-block: valve covers and bails, pushrod tubes, deflector tin, oil sump plate, oil pump cover, crank pulley, and bolt.

If you are building a Type 3 engine for a squareback or a fastback model, you need to use the correct tin underneath the cylinders. In the background is the standard deflector tin. In the foreground is the Type 3–only tin, otherwise known as the "cool tin."

Sump plate: These are the lowest thing on the car, so they usually take a beating. Replacements are inexpensive if yours is destroyed. Otherwise give them a quick blast and paint.

Oil pump cover: These tend to last forever, though they do get a little rusty on the outside. Also blast and paint this cover.

Crank pulley and bolt: Unless you are going to an aftermarket pulley or serpentine conversion, just blast and paint these items. It is easy to change out later if you change your mind.

Pushrod tubes: These were bought new. Most builders install them as is (bare metal). It's easy enough to wipe them down with some thinner and paint them. They tend to rust rather quickly.

Deflector tins: The tins are usually in good shape, so just blast and paint them.

Valve covers and bails: Sometimes these get so rusty that blasting them blows a hole right through them. Replacements are inexpensive. We just blasted and painted ours.

PREASSEMBLY

This chapter will cover a few things that don't take a lot of time but need to be done. Say you have a few minutes to kill in the shop or you are waiting for paint to dry. These things can be done at the beginning of the assembly process as long as you have the necessary components.

Preset the Flywheel Endplay

This is a great procedure for the first-time builder to perform. First timers usually don't have a large assortment of flywheel shims, and calculating the flywheel endplay ahead of time provides an opportunity to obtain the correct shims and keep the build on track.

Flywheel Endplay Components

- Flywheel
- Gland nut
- Rear main bearing
- Miscellaneous flywheel shims

Preset the Flywheel Endplay

1. Install the Bearing & Shims

With the crank clean and ready to go, place the new rear main bearing on the crank. Install two random flywheel endplay shims after that. They should be installed dry; no oil.

2. Oil the Gland Nut

Oil the flywheel gland nut before installing the flywheel.

3. Drive the Gland Nut

Drive the gland nut on with a 36-mm socket and 1/2-inch drive impact. Don't worry about torquing the nut precisely. You just need to know the flywheel is sitting flush with the end of the crank.

4. Measure the Gap

With a feeler gauge, measure the gap and write down that size. In our case, it was 0.013 and we are shooting for 0.006 maximum.

5. Remove the Gland Nut

Take the gland nut back off with the impact driver

7. Measure Two Test Shims

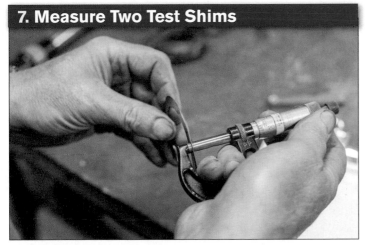

Our two random shims measured 0.023 with the micrometer and we had a 0.013 gap with the feeler gauge. That total is 0.036 and we need a 0.006 maximum gap (0.036 − 0.006 = 0.030). We need our recommended stack of three shims to total 0.030 inch. Shims come in six sizes ranging from 0.0095 inch to 0.015 inch. A variety pack retails for $15 to $20.

6. Remove the Flywheel

Remove the flywheel from the crank by gently hitting the backside of the flywheel with a dead blow hammer. Do this while holding the crank and flywheel just barely above the workbench.

Main Bearing Scribing Components

- 3-4 side of the engine case fresh from the machine shop
- Correct main bearings
- Main bearing dowel pins

Scribe the Main Bearings

Once the case has been line bored and the correct main bearings purchased, you can test fit them into the case. Install the tiny dowel pins into the case and carefully align the dowel pin hole in the bearing with the pin. Push the bearing in until it seats completely in the bearing journal.

Scribe the bearings with something sharp at the parting line of the case halves. This simple procedure could save you a ton of grief upon assembling the bottom end. The number one mistake first-time engine builders perform is not getting the dowel pin hole in the bearing lined up perfectly with the dowel pin in the case. Then they smash together, binding up the rotating assembly.

It's very easy to do; every builder has done it. You think the bearings are seated correctly on the dowel pins. You set the 1-2 side of the case on the 3-4 side. You begin to tighten the main bearing studs and the crankshaft is bound up tight. Hopefully you didn't tighten the main studs all the way before you noticed this issue.

If upon disassembly you find that the dowel has smashed into the bearing, missing the dowel pin hole, I have bad news for you. You'd be wise to order another set of bearings and try again. Next time, be extra careful to align the dowel pin with its hole.

Scribing the Bearings

1. Insert Dowel Pins

Insert the main bearing dowel pins into the 3-4 side of the case. The center main isn't necessary.

2. Pin Hole Offset

Each bearing has a dowel pin hole to locate the bearing correctly in the bearing journal. Notice it is off-set. The dowel pin hole always goes toward the flywheel.

3. Place Bearings In Bores

Place the bearings into their respective bores. Be sure that the bearing is seated flush and that the dowel pin and its hole line up. Just these three main bearings need to be marked. The center main bearing is split and won't need this procedure.

4. Scribe the Bearings

With a razor blade or pocket knife, scribe a line on both sides of the bearing where the two case halves split. Do this on all three completely solid bearings. During assembly, this simple step will ensure you have the bearings located correctly.

Gap the Piston Rings

Some piston and barrel kits come with the piston rings already installed on the pistons. If you are leery that the ring gaps are incorrect, you can remove them and double-check the gaps. From past experience, kits with the rings preinstalled have ring gaps within acceptable tolerances and can be used right out of the box. Since ours were manufactured from a different company than the piston and barrel company, we decided to check them before installing them on the pistons.

Gapping the Piston Rings

1. Install the Ring into the Cylinder

All of the piston rings need to be checked for end gap. Too much gap will cause a loss of compression and increased crankcase pressure. Too little gap will leave no clearance for the ring to expand once hot, which will possibly break the ring. Install a ring into a brand-new cylinder.

2. Level the Ring

Even out the ring in the bore by pushing it down with a piston. Level it with the bottom edge of the oil ring gap.

3. Measure the Gap

Measure the gap with a feeler gauge. Both the first ring and the second ring have the same gap specification: 0.012 to 0.018 inch.

4. File the Rings

If the gap is too tight, carefully file a slight amount off the end of the ring. Keep the file square to the end. It doesn't take much to make a drastic change, so take your time.

Install the Rings on the Pistons

Installing piston rings takes a bit of patience. If forced, they will break. The piston ring material is very brittle and breaks rather easily. If you think you lack the skills to install the piston rings without breaking them, contact someone with experience.

To get a feel for how much or little force they can withstand, grab the original pistons from the disassembly process and practice. Take the rings off and reinstall them on the pistons. If you break the old rings, no big deal. The old pistons and barrels were bound for the dumpster anyway.

Installing the Piston Rings

1. Install the Oil Ring Expander

Start with the oil ring expander. Place it in the bottom groove, making sure the ends butt up against each other and don't overlap.

2. Install the Bottom Oil Ring

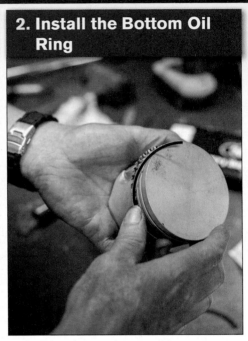

The lower oil ring goes on next, followed by the upper oil ring. These rings are fairly flexible and forgiving. Don't worry about indexing the ring gaps for now.

3. Install the Upper Oil Ring

The second ring gets started by pushing one end squarely into the groove. Gently push the ring into the groove with your thumb as you turn the piston. A small dot on the ring tells us which side goes toward the top of the piston.

4. Install the Second Ring

The second ring will get to a point where the ring catches on the upper edge of the piston. Carefully coax the ring over that edge a little at a time. The rings are made of cast iron and break very easily. Patience is key; don't force them.

5. Install the Top Ring

The top ring starts just like the second ring. Push the end square into the groove. Note the ring has a dot or is marked "TOP." This doesn't particularly mean that it is the top ring but that it needs to be installed with that side toward the face of the piston.

FINAL ASSEMBLY

Assembling an engine is the most enjoyable part of the entire engine rebuilding process. All the parts are clean and it's like assembling a model car or Legos kit. This will be the last time you see or touch some of these components, so be sure everything is ready to go. These many parts will become one machine, and to be successful we have to assemble them precisely and with painstaking detail.

Before we begin assembly, we need to clean up the work area. Clear the workbench of any unnecessary clutter. Different projects should be separated from the current one. You don't want to confuse items used for one thing when they are destined for a different one. If the parts have been sitting around for any length of time due to delays in your project, wash them again in the parts washer to remove any dust or debris. Even a shot of some brake cleaner and taking the air hose to all of the parts is a good idea. Lay out all of your clean components on rags or newspapers on the bench.

It is nearly impossible to assemble a VW engine without an engine stand. The 3-4 side of the engine case needs to be safely mounted on an engine stand that allows you to rotate it 360 degrees. This enables you to rotate the case as you install the pistons, barrels, and heads on either side of the case. You are basically building two 2-cylinder engines sharing one common crankshaft. The trouble is, each 2-cylinder engine is 180 degrees from the other one. Make the job easier and more enjoyable by purchasing the engine stand. In the end, you'll be glad you did.

Air-cooled VW engines are assembled differently than most automotive engines. The connecting rods are bolted on the crankshaft before the crank is installed in the engine case. The engine is assembled from the inside out, kind of like making a sandwich. You keep building on what is already assembled until you run out of parts. In this case, the valve covers will be the last thing installed in this chapter.

Crankshaft

Crankshaft preparation can take place as soon as all of the necessary components are available. Before installing our vise-mounted fixture, it's a good idea to thoroughly wash the crank (including the oil passages) using a rattail wire brush. If you've already done this, do it again. This dislodges any stray particles that may have been missed during initial cleanup after machine shop work. Once the cleanup is accomplished, measure the crankshaft journals one more time to confirm proper bearing sizing before installation.

A number of items are installed on the pulley end of the crankshaft. Two main bearings and the gears for the cam and distributor drive are among the clips and keys. Our engine case was line bored 0.020 inch (0.50 mm) oversize and our crankshaft is ground 0.010 inch (0.25 mm) undersize, so the corresponding main bearings were purchased.

Main Bearings

From the box of main bearings, we pulled two bearings out to install on the crank at this time. They are the only two bearings that look similar but are drastically different in size. The larger one goes on first.

First, pull out your hot plate, plug it in, and turn it all the way up. Set the larger crank gear on the hot plate with a drop of oil on it.

Now back to the bearings. Apply plenty of assembly lube to the larger bearing and install it first. Make sure the dowel pin hole is facing the flywheel end of the crank.

CHAPTER 7

Tech Tip

Use an abundance of assembly lube on all of these components. The reason for this is simple; even with the best-laid plans, engines sometimes go for months and even years before they are installed in the vehicle. Bearings and journals need plenty of protection during this time. And when it's time to fire it up, the only protection on these freshly machined surfaces is this assembly lube until the oil system gets primed. ■

Check to see if the cam gear is warmed up; the oil smoking is an indicator it's plenty warm. Grab it with a channellock pliers. Make sure the chamfer is going on first and the two dots for the cam timing are visible toward the end. Line the keyway up with the key on the crank.

Next comes the gear spacer. The brass distributor drive gear can get warmed up a little but shouldn't be as tight going on as the cam gear. Next is the clip, which can be a bear. If everything is on correctly and seated properly, the clip groove should be completely uncovered. A

good, heavy-duty snap ring pliers is key for your success.

Next, the smaller main bearing is installed. Again, apply plenty of lube. After that is a disk-shaped washer called an oil slinger. This is a very important item, and if it is omitted or installed backward, it can cause a lot of grief. Rather than have an oil seal at the pulley end of the crank, VW engineers designed a way for oil to get slung off the crank using centrifugal force. The oil slinger is dished toward the end of the crank, and oil spins off it before making its way down the crank and out of the case in the form of an oil leak.

This oil slinger and the crank pulley design (with a screw feature to screw the oil back into the case) are the two ways the engineers eliminated the need for a seal on the pulley end of the engine. After the oil slinger is installed correctly, the pulley key can be pushed into the keyway with pliers.

Connecting Rods

The connecting rods can be installed as long as you have the correct bearings at this time. Our crankshaft was ground 0.010 inch (0.25

mm) undersize on the mains and rods. The connecting rods have a right and wrong way of going on the crankshaft. The rod caps at the big end of the rod can also be installed incorrectly.

Factory connecting rods have a bump midway up the beam. This bump must be facing up as the engine is installed in the vehicle. The way we have the crankshaft positioned in the fixture clamped in the vise, it may be a little hard to visualize which way is up. Just remember the rod journal closest to the flywheel end of the crankshaft is cylinder number-3. Number-3 is on the left side of the engine and closest to the front of the vehicle.

The factory also stamped numbers on both halves of the rod at the split of the big end. These numbers should both be on the same side of the rod as the bump on the beam. Lastly, the tangs that hold the bearings in are always opposite the bump. That's the feature used for our build. The aftermarket connecting rods don't have the bump or numbers stamped in them. All we have to go by is the tangs, which should both be at the bottom of the engine when installed in the vehicle.

Crankshaft Preparations

Special Tool
1. Install the Crank Holder

Install the special crank holder into the end of the crankshaft.

2. Clamp in Vise

Clamp the holder in a vise.

3. Heat Up Cam Drive

Heat up the cam drive gear on the hot plate. Once the drop of assembly oil starts to smoke, it's plenty hot.

4. Install the Main Bearing

Install the new main bearing while the gear is heating up. It's the only one in the set that looks like this one. Note the dowel pin hole is offset toward the flywheel end. Use plenty of assembly lube.

5. Install the Cam Drive Gear

Install the hot cam drive gear. Make sure the large inner chamfer faces down and the timing dots face up.

6. Install the Distributor Drive Gear

Once you pull the cam gear off the hot plate, set the brass distributor drive gear on the hot plate. Install the gear spacer and warmed up distributor drive gear.

7. Install the Snap Ring

Install the snap ring with the heavy-duty snap ring pliers. Be careful because things may be hot yet. If you can't see the groove for the snap ring, then something isn't seated down all the way. Double-check to make sure the cam gear is on correctly.

8. Install the Nose Main Bearing

Install the nose main bearing. Again use plenty of assembly lube and make sure the dowel pin hole is offset toward the flywheel end of the crank.

9. Install the Oil Slinger

Install the oil slinger. Make sure it is dished up toward the pulley end of the crankshaft, as shown. Install the pulley key with a channellock pliers; it should push right in.

10. Install the Rod Bearings

Insert all of the rod bearings into the rods. Our crank is ground 0.010 under-size, so we are using −0.25-mm rod bearings. Use plenty of assembly lube.

11. Install the Nuts

Install the nuts with the turned down part against the rod. Be sure not to flip the rod cap around. The tangs that hold the bearings should meet up. On factory rods, make sure the numbers stamped in them are together.

12. Inspect the Bumps

Pay attention to the bump on the beam of the rod. It needs to face up when the engine is installed in the car. The view shown is from the bottom of the engine when installed, so the bumps are hidden.

13. Torque the Nuts

If you are using factory rods, oil the threads and torque the nuts to 24-25 ft-lbs using a 14-mm socket. These replacement rods have a different torque spec, so we followed the instructions that came with them.

14. Inspect the Assembly

This is the top view of how your assembly should look when installed. Note the position of the rods and where the bumps on the beams are.

Engine Case Preparation

Each half of the engine case will need some attention before the rotating assembly and lower valvetrain components are installed. The two mating surfaces need to be clean and void of any assembly lube or oil of any sort. A quick pass with a flat file over both of them will remove any nicks or bumps they may have incurred. Spray both halves down with brake cleaner and blow them off thoroughly with compressed air.

Bolt the 3-4 engine case side onto the yoke of the special VW engine stand. You will need to find some mounting hardware for this step. All engines with doghouse-style oil coolers will have a 10-mm threaded boss pressed into the upper engine mount location. A 10-mm bolt that is 40-mm long will work there.

The lower engine mount is a 10-mm stud screwed into the case. It will be necessary to put a spacer approximately 1½ inches long with a 7/16- to 1/2-inch hole to fit over the stud and install a nut. The stud isn't threaded all the way, so a nut can't be installed without a spacer.

The spacer must be small enough as to not interfere with the flywheel once installed.

Install the cleaned up head studs in both halves of the engine case. The longer ones go along the lower portion of the barrel openings, near the pushrod tube holes. The ninth long stud goes in the deep hole at cylinder number-3 and should stick out the same amount as the opposite corner at cylinder number-4.

The shortest studs are at the upper middle on both sides. This holds true for all the dual-port 1,600 engines, no matter if they are 8-mm head studs or 10-mm head studs. If your engine is a single-port or a 40-hp model, your head studs will only be two different lengths: eight long and eight short.

Long ones are installed along the lower portion of the barrel openings, and the shorter ones go along the upper portion. Some of the stud holes go straight through into the interior of the case. An oil leak will develop if these studs are installed without some sort of sealer. Aviation gasket sealer works, as does oil-resistant RTV. Don't use Loctite

or any kind of thread locker on any of the head studs. The next guy that rebuilds this engine will thank you.

All the little dowel pins that locate the main bearings can be installed now. There are four in the 3-4 side and one in the 1-2 side. The split main bearings can be installed in both halves. The six seals that slide over the six main bearing studs need to be installed. The cam bearings can be installed in both halves.

The lifters are all the same and can be pushed into each half with plenty of assembly lube. We chose reground, genuine VW lifters instead of new aftermarket lifters. The quality of the cheaper aftermarket ones have been hit or miss. Hold the lifters into the 1-2 side of the engine case with special clips. The clips will hold the lifters in place while you flip the case over and set it down over the 3-4 side. The lifters falling out of their bores and bouncing around on the floor or the inside of the 3-4 side will not make your day.

Aside from the sealer to seal the two halves together, these assembles are almost ready to become one again.

Preparation of the 3-4 Side of the Engine Case

1. Inspect the Head Studs

These are the 16 head studs all cleaned up, ready to install. Notice there are nine long ones, three medium length ones, and four shorter ones.

2. Install the Long Head Studs

Bolt the 3-4 side of the engine case to your engine stand. Install the longest head studs along the bottom, nearest the pushrod holes. Use gasket sealer on the threads, as some of the stud holes are clear through.

3. Inspect the Head Stud Lengths

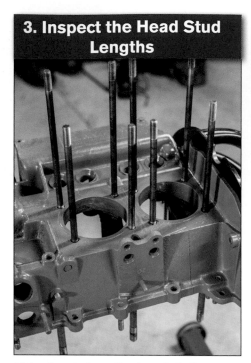

This view is how the head stud lengths should end up. Long ones along the bottom. The odd long stud gets installed in the deep hole at the number-3 cylinder, the two short ones are in the upper center holes, and the one medium length stud is at the outside of the number-4 cylinder.

4. Install the Cam Bearings

There are six cam bearing halves, and all of them have a specific location. Install the only one with thrust surfaces in the journal nearest the cam gear, the widest ones in the middle, and the narrow ones on the end.

5. Inspect the Studs & Cam Bearings

This is a finished view of everything you need in the 3-4 side of the case: four main bearing dowels (one is under the center main bearing half), one center main bearing, six main stud rubber seals, three cam bearings, and four lifters with plenty of assembly lube.

Preparation of the 1-2 Side of the Engine Case

1. Install the Remaing Studs

Install the remaining eight head studs in the 1-2 side of the engine case. There are four long ones along the bottom, two medium length ones on the upper corners, and two shorter ones in the center. Again, apply some gasket sealer to the threads because some of the stud holes go clear through.

2. Inspect All Components

Here are the necessary components installed in the 1-2 case half: one bearing dowel pin under one center main bearing, three cam bearings, and four lifters with plenty of assembly lube.

Special Tool

3. Install the Lifter Clips

These handy clips keep the lifters in their bores while you flip this half over and set it on the other half.

Join the Two Halves

The rotating assembly and cam need to be placed into the 3-4 side engine case that is bolted onto the engine stand. Remove the crankshaft from the vise and unscrew the vise fixture. Slide the flanged main bearing onto the flywheel end of the crankshaft. Use plenty of assembly lube. Remember, the dowel pin hole is toward the flywheel.

With all the little things ready to go with the case halves, now is the time to apply the sealer. Aviation gasket sealer is used by the majority of engine builders. There are other similar products on the market. Honda-bond and some Permatex products are specifically made for joining two bare metal surfaces together.

The main thing the gasket sealer needs to do is seal the two surfaces without adding dimension between them. In other words, it needs to almost completely squeeze out from between them, leaving only the bare minimum of product to seal them

together. Many a home VW engine rebuild has been ruined by using silicone sealer, RTV, or some other type of "gasket maker" as a sealing agent between the two halves. In reality, the two case halves were never completely together. That changes all the clearances and the engine will soon fail.

Some preliminary adjustments to the connecting rod positions and main bearing positions should be made before you grab the crankshaft assembly and put it into the 3-4 side of the case. You are going to grab the assembly by the wrist pin ends of number-1 and number-2 connecting rods, letting the number-3 and number-4 rod droop down.

Index the three main bearings on the crank so when you lift it up by the number-1 and number-2 rods, the dowel pin holes are facing straight down. That will save some time trying to align the holes onto the dowel pins. Not having the dowel pin in line with the hole in the bearings is the number-one problem amateur VW engine builders face. Even cocky, sea-

soned veterans will catch themselves thinking everything is lined up when it isn't. Scribe the individual bearings with a line at the case parting line with something sharp beforehand.

With the crankshaft assembly placed correctly in the 3-4 case half, the cam goes in next. Rotate the crankshaft assembly by pushing and pulling on the number-1 and number-2 connecting rods until you see the two dots on the crank gear. Put the dots somewhere near the cam journals. Apply assembly lube to all the cam bearings and lifter faces on both halves. Mesh the cam gear with the crank gear, making sure the one dot on the cam gear is between the two dots on the crank gear. Keeping the gears meshed, place the cam into the bearings.

The cam plug can be installed now with a nice coating of gasket sealer. It doesn't matter which way it goes in for a manual transmission. For an automatic or auto-stick transmission vehicle, the solid end of the plug should face inward or interference with the flexplate could occur. The predetermined amount of flywheel endplay shims (see chapter 6) can go against the thrust bearing with a little assembly oil.

Rather than installing the rear main seal once the case halves are bolted together, save yourself the hassle and install it into one half now. Just set it in place after the flywheel shims are installed. Apply some assembly oil to the lip of the seal where it rides on the flywheel to prevent premature wear of the seal upon startup.

Grab the 1-2 side engine case half by the head studs and locate the six main studs by looking through the holes they come up through. Gently lower the case half until it stops from its own weight. The only things that should be stopping the halves from joining completely are the two dowels that align the two halves. These 8-mm dowel pins are at opposite corners of the case: one is below the oil pump opening and the other one is at the top of the case where the case bolts to the transmission. By squeezing these points with a channellock pliers, the two halves should join together easily.

Install the six special main stud washers and some assembly oil on the threads. Install the six nuts with a 17-mm socket, but don't tighten them. Keep them hand tight for now. Install the hand crank you made into the pulley end of the crankshaft.

Rotate the crankshaft a couple degrees; it should move very easily. Tighten the two center mains slightly while rotating the crankshaft back and forth a little bit. It should stay loose. If it binds up tight, you have a problem.

If it loosens up with the nuts backed off, then a dowel pin isn't in its hole and is pinching the bearing against the crankshaft. You have no choice but to pull it apart and see where you went wrong. Clean up the sealer and start again.

Once you get the six mains torqued to 25 ft-lbs, install the perimeter case wave washers and nuts. Torque to 14 ft-lbs with a 13-mm socket.

Marriage of the Engine Case Halves

1. Install the Main Thrust Bearing

Install the thrust main bearing with plenty of assembly lube. Note the dowel pin hole is offset toward the flywheel. Add some assembly lube to the center main bearing in the 3-4 half.

4. Check the Dowel Pin Alignment

2. Lower the Crank Assembly Into Case

Grab the crankshaft assembly by the number-1 and number-2 connecting rods. Make sure the dowel pin holes in the bearings are facing down. Set the crankshaft assembly into the case as best you can.

3. Align the Dowel Pin Hole

While lifting up on the crankshaft, align the dowel pin hole in the bearing with the dowel pin sticking up out of the case.

All three solid main bearings need to be sitting directly on the dowel pins. This is the reason we pre-fit those bearings individually and scribed lines at the split in the case halves.

Critical Inspection

5. Inspect the Scribe Lines

Double-check with a flashlight that the scribe lines you put on the bearings are directly on the split point of the case halves. This is a very crucial step.

6. Apply Assembly Lube

Apply assembly lube to the cam bearings and the lifter faces of the 3-4 side.

7. Install the Cam

Install the cam, taking note of the timing marks. The one dimple in the cam gear has to go between the two dimples in the crank gear.

8. Lube the Cam Lobes

Carefully apply more assembly lube to the cam lobes, trying not to disturb the cam timing.

9. Lube the Cam Bearings

Apply assembly lube to the cam bearings, lifter faces, and center main bearing in the 1-2 side of the engine case.

10. Install the Cam Plug

Apply gasket sealer to the cam plug groove in the 3-4 side. Push the cam plug into the groove. Apply more gasket sealer to the cam plug once it is pushed in.

11. Lube & Install the Shims

Apply a thin coat of assembly oil to each of the three flywheel shims we predetermined in chapter 6. Install them on the end of the crankshaft.

12. Install the Rear Main Seal

Install the new rear main seal into the 3-4 half of the case. No sealer is necessary. Be sure to apply some assembly oil to the inner seal lip.

13. Install the Case Sealer

Apply gasket sealer to the 1-2 side of the engine case along the outer perimeter edges. It is easier to apply to this half rather than the 3-4 side with everything installed and in the way.

Pro Tip

14. Stand Up The 1-2 Rods

PRO TIP *A trick here to keep the number-1 and number-2 connecting rods standing up long enough for you to drop on the 1-2 side of the case. Squeeze the two rods together right before.*

15. Marry the Case Sides

Grab the 1-2 case half by the head studs and gently drop it onto the 3-4 side. Look through the main stud holes to line them up.

16. Clean & Organize Hardware

Our case hardware is all cleaned up and separated into sizes in our secondhand muffin tin. This is a great way to be more efficient.

17. Install the Washers

Place the six washers on the main studs and apply some assembly oil on the threads.

18. Hand Tighten the Studs

Cinch down the six main studs hand tight with a 17-mm socket. Make sure you can easily spin over the crank. Any binding now will indicate the dowel pins in the main bearings are not located correctly.

19. Torque the Studs to Spec

Torque the six main studs to 25 ft-lbs using the correct torque pattern. Double-check to make sure the crank spins freely. Problems now indicate the case will have to be split back apart to resolve the issue.

20. Torque Pattern

Torque the six main studs using the pattern shown here.

21. Apply Gasket Sealer

Apply gasket sealer to the two studs on either side of the rear cam bearing. These tend to seep oil over time.

22. Install Perimeter Wave Washer & Nuts

Install all the perimeter 8-mm wave washers and nuts. Eleven are washers and nuts and three are bolts with washers and nuts. The one longer stud on the top of the case is to mount the intake manifold center section, and it doesn't get hardware yet.

23. Torque Perimeter Nuts

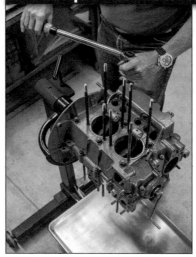

Torque all of the perimeter nuts to 14 ft-lbs using a 13-mm socket.

Flywheel

Our original flywheel had seen better days. It needed to be reground and the rear main sealing surface was questionable, so we opted to replace it. The replacement only cost $10 more than what it cost to regrind an old one.

A fresh flywheel O-ring from the gasket kit was installed. Use a little assembly oil on the thrust face and seal surface. There's a thin tin gasket in the gasket kit that looks like it should go on before the flywheel goes on but it isn't used in this case. The older models without the O-ring in the flywheel get that. The flywheel isn't indexed to the crankshaft in any way.

The four dowel pins in the crank are equally spaced. Our gland nut had seen better days as well, so we got a new one. Add a couple drops of assembly oil on the threads and washer. Get it as tight as you can by hand. Install the flywheel lock to stop the crankshaft from spinning. At this point, you need to torque the flywheel, but you also need to stop the engine from spinning in the engine stand. The gland nut gets torqued to 253 ft-lbs and everything is going to want to spin!

A length of 2x4 wedged between the arms of the engine stand yoke will stop it from spinning. Another option is to use the piece of angle iron we made for removing the flywheel initially.

Bolt that to the flywheel and you don't have to use the flywheel lock. Either way, you need the flywheel to not turn under the force of 253 ft-lbs. A 3/4-inch drive 1⁷/₁₆-inch socket and extension were used through the center of the engine stand yoke. We actually set our torque wrench to 253+10 (4 percent) or 263 ft-lbs to compensate for the torque wrench error of +/- 4 percent. We wanted to be on the plus side.

Once torqued, the flywheel back-lash or endplay must be checked. Most builders use a dial indicator and a magnetic base mounted on the flywheel. Push the flywheel in as hard as you can and set the indicator to zero. Now push the flywheel out and note the reading. The acceptable range is 0.003 to 0.005 inch. Most builders agree that closer to 0.005 inch is better than not. If it is less than 0.002 inch or more than 0.005 inch, the flywheel needs to come back off and the shims changed out to get the correct total shim thickness to achieve endplay in the acceptable range.

By using an indicator holder that mounts to the engine case not only can you check endplay but you can also check flywheel wobble or run-out. The factory spec for run-out is 0.012 inch, measured at the center of the clutch surface. Check run-out by rotating the crankshaft with the hand crank while keeping constant inward pressure. The total range of motion should be less than 0.012 inch on the indicator.

Installing the Flywheel

1. Install the O-Ring

Install a new flywheel O-ring into the flywheel.

2. Apply Assembly Oil

Apply assembly oil to the thrust surface and the rear main sealing surface of the flywheel.

3. Attach the Flywheel

Install the flywheel onto the crankshaft dowel pins. The orientation doesn't matter, the pin locations are symmetrical.

4. Lube the Gland Nut

Apply assembly lube to the gland nut threads and both sides of the washer.

Special Tool

5. Install the Flywheel Lock Tool

Install the flywheel lock tool and start the gland nut into the end of the crankshaft.

6. Secure the Engine Stand

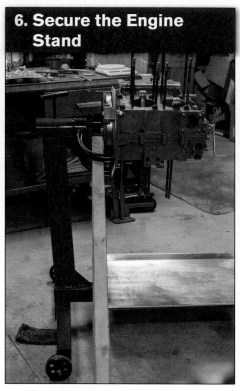

Place a 2x4 between the arms of the engine stand yoke. It needs to be approximately 6 feet long. This is to keep the engine stand from flipping over while you torque the gland nut.

7. Torque the Gland Nut

Set your giant 3/4-inch drive torque wrench to 253 ft-lbs. Torque the gland nut using a 1$^7/_{16}$-inch 3/4-inch drive socket. We went directly through the yoke of the engine stand with a 3/4-inch drive extension.

8. Install the Dial Indicator

Mount a dial indicator on the engine case. Put the point at the center of the clutch surface and zero out the dial while pushing out the crank from the other end.

9. Check End Play

Push the flywheel in and out while you watch the needle. You are shooting for 0.003- to 0.005-inch travel. Ours is perfect at 0.004 inch. Now crank the engine over while pushing in on the end on the crank to measure flywheel run-out. Maximum run-out is 0.012 inch; ours was 0.005 inch.

Piston

In the previous chapter, we checked all the piston ring gaps and installed the rings on all four pistons. Pistons in a VW engine are floating pin pistons. The wrist pin is kept in place with clips on either end of the wrist pin hole.

These clips can be tricky to install, so take your time and make sure the clip is seated completely in the groove. If the clip isn't completely in the groove, you probably won't find out for quite some time. The wrist pin will eventually push the clip over to the cylinder wall and start digging a groove into the cylinder wall. Your fresh rebuild will suffer from oil-burning, low compression, and increased blowby. All from one little clip that didn't seat properly.

Install one clip in each piston while you can still hold them in your hand. Take a good look and be sure the clip is in place. The piston and barrel kits come with exactly the right amount of parts to complete the job. Count out eight clips and set them out where you can be sure of the amount. If you finish installing all four pistons and barrels and you see a wrist pin clip still on the bench, you got lucky and saved your engine from disaster. Start removing parts until you find the culprit.

VW engineers also designed the pistons with an offset to the wrist pins. This means the wrist pin location isn't exactly on center; it's shifted slightly (1.5 mm) to put less stress on the piston skirts and decrease engine noise. This is the reason for the arrows on the piston faces.

It makes no difference what piston goes where as long as the arrow points toward the flywheel. That is critical. The engine will still run with the pistons installed incorrectly but the original intention of the designers (longer life and quieter operation) will be greatly diminished. Just pay attention and install the pistons the way the engineers designed them and everyone will be happy.

Piston Installation

1. Install the Clip

Install one wrist pin clip into each of the pistons.

2. Lube the Rods

Lube the small end of the connecting rods with assembly oil.

3. Install the Wrist Pin

Push the wrist pin into the piston and rods. Take note of the way the arrow is pointing on the face of the piston. The arrow always points toward the flywheel. The wrist pin hole in the piston is offset to put less load on the piston skirts and increase ring and barrel life.

4. Install the Second Clip

Install the other clip in the piston and make sure the clip is seated completely in the groove of the piston.

5. Repeat for the Other Pistons

Repeat the process with all the pistons, making sure that the arrow on the face of the piston always points toward the flywheel.

Barrel

V-8 engines are assembled by pushing the piston (and most of the connecting rod) down the cylinder bore with oil rings first. VW engines are assembled the opposite way: the barrel or bore is pushed down over the top of the piston with the compression ring first. Oil control rings are much more forgiving than brittle compression rings. Extra care must be taken not to damage the compression rings while tapping the barrel over the piston.

Paper gaskets seal the bottom of the barrel to the engine case. A light coat of gasket sealer keeps them in place while installing the barrels. A little assembly oil coating on the inside of the barrel helps with initial startup, and if the engine is going to sit for some time before actually firing it up, the coat of oil will stop the bore surfaces from rusting and becoming a problem before you even get the engine in the vehicle.

Right before installing the piston ring compressor is a good time to index the rings. Follow the piston ring indexing diagram in the appendix. The oil ring expander gap is always at 12 o'clock with the oil control rings on either side of that at roughly 11

and 1 o'clock. The reasoning for that is the oil gets splashed up into the barrels while the engine is running. Once the engine stops, the oil will run toward the bottom of the barrel and can bleed past the oil rings and leak into the combustion side of the piston. This will cause oil to burn and smoke at startup.

The two compression ring gaps need to be at roughly 8 and 4 o'clock. This will keep them far enough away from the oil control ring gaps and more than 90 degrees apart from each other, per the manufacturer's recommendations. The pistons do the least amount of rocking at the center line of the wrist pin. Having the ring gaps near that helps the solid part of the ring seal better at the point it does the most rocking back and forth.

Compress the rings with a good ratcheting ring compressor. Place the barrel down over the piston and tap down on it with the entire face of your hand. It should move a little with each hit. Keep a little extra tension on the ring compressor as well. If you hit it with your hand and it doesn't move, stop! Make sure none of the rings popped out of the ring compressor and got caught on the barrel.

The key here is to try not to break the piston ring. Rotate the crankshaft

in the case 180 degrees and do the other piston. It doesn't matter what side of the engine you start on, they all need to be put on. After the barrels is a much overlooked and forgotten item: the deflector tin. This small piece of sheet metal is key to keeping the cooling even across the entire barrel surface. Without it, the barrels run hot along the bottoms and are prone to warping and getting out of round from hot distortion.

Out-of-round barrels have loss of compression and power. They also have increased oil consumption and blowby. This little piece of sheet metal has a big job, and if forgotten, it is impossible to install later without pulling out the cylinder heads and pushrod tubes.

After the deflector tin comes the pushrod tubes. These come in stainless steel or aluminum and are internally or externally spring-loaded. On a stock rebuild, it's hard to beat the stock crush-to-fit steel ones. The only downside to them is they rust rather quickly. The solution to that is a quick paint job.

While installing them, be sure to index the weld seam upward. Once in a blue moon a pushrod tube will crack and split along that seam and cause an oil leak. With the seam facing upward, the leak usually isn't bad at all and almost livable. With a leaking seam facing down, the oil leak will be annoying and bad enough that it will need attention as soon as possible.

A word about pushrod tube seals. Some gasket kits have red tube seals that tend to get very hard very fast. They will leak much sooner than they should. The good seals are white and stay soft for a very long time. That being said, a nice ring of oil-resistant silicone sealer is still a good idea for an oil-tight seal against the head and case.

Barrel Installation

1. Clean the Barrel

Clean the lower sealing surface of the barrel with brake cleaner.

2. Apply the Sealer

Apply a coat of gasket sealer.

3. Install the Gasket

Install a gasket from the gasket kit.

4. Apply More Sealer

Apply another coat of gasket sealer.

5. Index the Rings

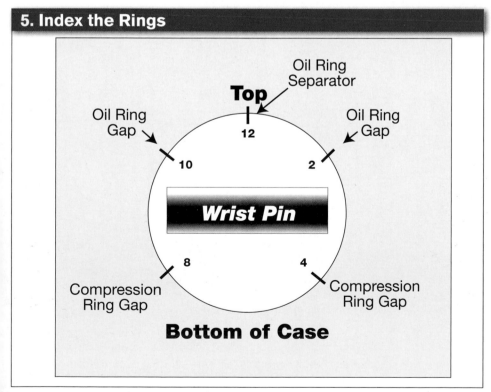

Index the piston rings as shown in this diagram.

6. Compress the Piston Rings

Compress the piston rings with a piston ring compressor.

7. Apply Oil

Apply a thin coat of assembly oil to the inside of the barrel.

Critical Inspection

10. Measure the Deck Height

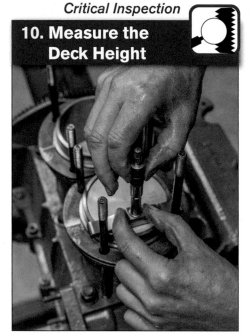

If you are calculating compression ratio, now is a good time to measure the deck height. Rotate the engine until the crank pulley key is exactly at 9 o'clock when the engine is as it would be installed in the car. Measure the deck height with a depth micrometer. Make sure the barrel is pushed flush with the case. Ours measured 0.085 inch. Write this down to use in a compression calculator formula.

8. Push Barrel onto Piston

Push the barrel onto the piston. Be careful not to catch the piston rings on the edge of the barrel.

11. Install the Deflector Tin

This important step is easily overlooked. The deflector tin must be installed before the heads are bolted on.

9. Repeat on the Other Side

Rotate the crank 180 degrees. Repeat the process, compressing the rings and installing the other barrel.

The deflector tin clips onto the lower center head studs. They are head stud size specific: 8-mm or 10-mm.

How to Calculate Compression Ratio

By far, the easiest way to calculate your compression ratio is by going online and searching "Compression Ratio Calculator" then plugging in the numbers. The calculator then spits out the compression ratio. But what are the numbers the calculator needs? Most formulas need to know the bore, stroke, deck height, combustion chamber volume, and number of cylinders. Some even let you pick mm or inches for the data. It couldn't get easier!

For those of you who like to do things the hard way, here is a simple formula. All you need is the right information and a basic, pocket calculator.

A = Volume of one cylinder in cc
B = Volume of the deck height in cc
C = Volume of the combustion chamber in cc

$$(A + B + C) / (B + C)$$

Let's break down the three elements required in the formula.

A: The engine displacement in cc:
Engine displacement is bore area x stroke x number of cylinders.

The four sizes covered in this manual are:
77-mm bore x 64-mm stroke or 1,192 cc (1,200 or 40 hp)
77-mm bore x 69-mm stroke or 1,286 cc (1,300)
83-mm bore x 69-mm stroke or 1,493 cc (1,500)
85.5-mm bore x 69-mm stroke or 1,584 cc (1,600)

Ours is a 1,600 or 1,584 cc to be exact. Divide that by four (number of cylinders) = 396 cc = A

B: The deck height volume in cc:
This one is a little tricky to figure out.
First, you need to know the deck height in mm.

Ours was 0.085 inch x 25.4 = 2.16 mm. The formula looks like this:

$$\frac{((bore \times bore) \times deck\ height) \times 0.00314}{4}$$

$$\frac{((85.5 \times 85.5) \times 2.16) \times .00314}{4}$$

$$\frac{(7310.25 \times 2.16) \times .00314}{4}$$

$$\frac{49.581}{4} = 12.4\ cc$$

C: The combustion chamber volume in cc:
In chapter 5, we went through the process of figuring out how to find the volume of the combustion chamber or "CC" of the head. Ours ended up 52 cc, which is in the range of what a stock head should be. A volume of between 52 and 53 cubic centimeters. So 52cc = C.

With these numbers, let's figure out our compression ratio.

$$396 + 12.40 + 52 = 460.4$$
$$12.40 + 52 = 64.4$$
460.4 / 64.4 = 7.149:1 compression ratio, which is very close to the factory specification of 7.3:1.

The factory also recommended a fuel-octane minimum of 91, and with our lower compression ratio, the fuel octane won't be an issue. It will run fine on less expensive 87-octane fuel. ∎

Cylinder Heads

Stock VW air-cooled engines don't use head gaskets. The aluminum head is held on with eight long head studs per head and seal well in a huge range of temperatures and conditions. With that being said, head stud torque is critical to the longevity of the engine.

The heads are torqued down using two simple diagrams. The first pattern is very low torque (5 ft-lbs) and is basically to crush the pushrod tubes and get them set in place. The second pattern is just like any other head torque pattern. It starts from the center and works its way to the outer studs, alternating side to side. Obviously, the 10-mm head studs have a higher torque than the 8-mm head studs: 23 ft-lbs versus 18 ft-lbs.

1. Start the Head on Studs

Rotate the engine in the engine stand as shown and get the cylinder head started onto the head studs.

2. Seal the Pushrod Tubes

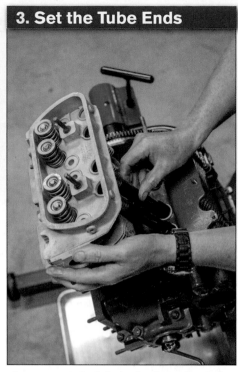

Apply a thin coat of silicone sealer to the pushrod tube seals. Note the dark line that runs the length of the tube. That is the seam that is welded together. Make sure that seam faces up when the engine is installed in the car. Periodically, that seam splits and if it faces down you will have a pesky oil leak.

3. Set the Tube Ends

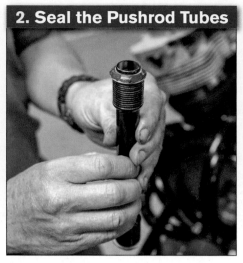

Set one end of the tube into the case side and let it rest against the cutouts in the head.

4. Install the Head

Install the cylinder head while aligning the pushrod tubes into their respective spots.

5. Install the Head Stud Washers

Install the special head stud washers and apply assembly oil to the threads.

6. Install the Nuts

Install all the special head stud nuts with a 15-mm socket to the point they just touch the washers.

7. Perform the Initial Torque

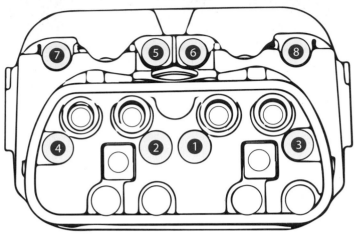

Torque the head to 7 ft-lbs using the sequence shown. This procedure is to properly crush the pushrod tubes.

8. Final Head Torque

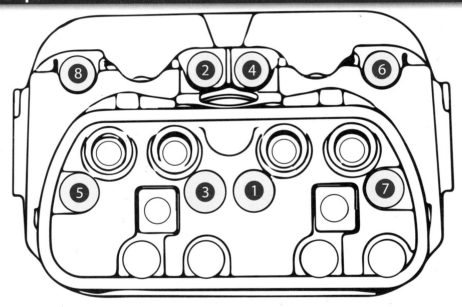

The final head torque sequence is different from the initial torque sequence. It basically moves from the center out. Torque the 10-mm head studs to 23 ft-lbs (8-mm head studs get torqued to 18 ft-lbs).

Oil Pressure Relief Valves

Early VW engine designs had issues with the oil coolers bursting upon startup in very cold temperatures. The thicker oil increased the oil pressure to more than the oil coolers were designed to withstand and they'd burst. To combat this issue, Volkswagen designed an oil pressure relief system that bypassed the oil cooler when the oil was cold and thick. The new system sent oil directly to the bearings, speeding up the warm-up process. Once warm, the thinner oil had less pressure and went to the cooler.

Later on, VW designers added a second oil pressure relief valve at the end of the oiling system to bypass excessive oil pressure directly back into the sump. All 1,600-cc dual-port engines have dual oil relief cases. Everything else will have single oil relief cases.

1. Inspect the Parts

The components of the oil pressure relief system are shown. The piston with the groove and the long spring go in the hole closest to the pulley. The plain piston and short spring go in the hole closest to the flywheel.

2. Install the Piston & Short Spring

Drop the plain piston and short spring into the hole near the flywheel.

3. Add Gasket Sealer

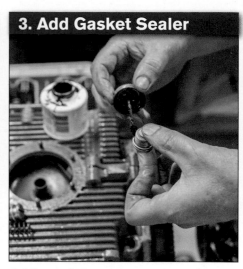

Adding a little gasket sealer on the new crush gasket from the gasket kit is a good idea.

4. Tighten the Plug

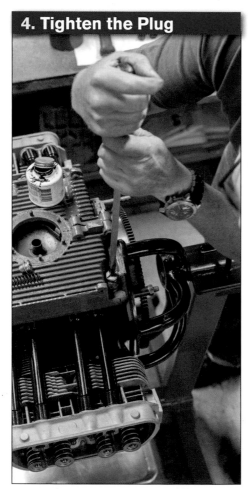

The big screwdriver is the tool of choice here. There is no real torque spec for this plug, but get it as tight as you can with the screwdriver.

5. Install the Long Piston & Spring

The grooved piston and longer spring go in the other hole. Note the spring sticks out much farther than the other spring did.

7. Apply Final Torque

6. Install the Second Plug

Again apply some sealer to the crush washer and compress the spring with the plug. Get it started by hand while holding the spring down. Finish tightening with the big screwdriver.

The plugs are nearly impossible to torque accurately. The special tool for removing them can be used with more leverage, then tighten them as well.

Full-Flow Filter Oil Pump Upgrade

The oil pump of our donor engine checked out as good. We could have cleaned it up and reinstalled it, but we decided to go for an upgrade. Full-flow filter pumps, such as EMPI part number 9207, are a great upgrade over the stock pump, plus you get the added bonus of a standard, spin-on oil filter. It has 32-mm gears providing added oil pressure, increased oil volume, and oil cooling. The oil filter cleans the oil 100 percent of the time.

It must be noted that these filter pumps will not fit the stock VW Beetle muffler. They fit just about every aftermarket exhaust available. Also it must be noted that these pumps will not fit Type 2 buses or Type 3 fastback and squareback vehicles. The rear engine hanger interferes with the filter housing.

Our engine is destined for the vehicle it came out of: a 1970 VW Baja Bug. It will be fitted with an off-road-style exhaust, so fitment will not be an issue.

1. Inspect New Parts

We are installing a new oil pump with a built-in oil filter housing. These filter pumps can only be used in Type 1 applications without the factory muffler. EMPI part number 9207 is correct for our application with a dished cam gear. This pump has a common oil filter available at any auto parts store. Replacement oil filters are listed in the tune-up specs in the back of the book. (Photo Courtesy EMPI Inc.)

2. Install the Oil Pump Studs

With all the oil pump studs removed, reinstall one of them a thread or two. It doesn't matter which one.

3. Apply Gasket Sealer

Apply gasket sealer to the pump housing and to the gasket supplied in the oil pump kit.

4. Tap the Housing into the Opening

Tap the pump housing into the opening. Note that it can only go in one way. The idler gear toward the bottom of the case and the driven gear hole line up with the end of the cam.

5. Test Fit for Clearance

At this point, test fit the filter housing for clearance. It touches a portion of the case that is usually machined away because this original case casting could've been used in a Type 2 bus application where a rear engine support is mounted to the case.

6. Adjust Clearance for the Pump Body

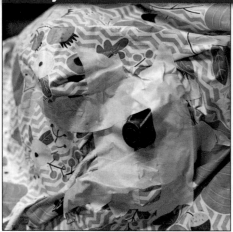
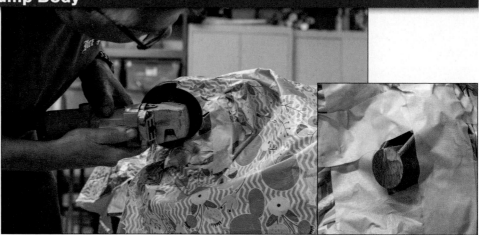

To remedy the clearance issue, a slight amount of material has to be removed from the already partially assembled engine. The engine was covered completely with plastic (in this case a nice holiday tablecloth). The area we have to clearance is exposed and taped up with masking tape. The angle grinder was used to remove just enough material.

7. Apply Sealer to the Housing

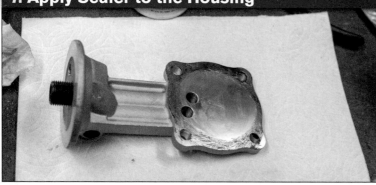

Apply a thin coat of gasket sealer to the filter housing. Don't get carried away.

8. Apply Sealer to the Gasket

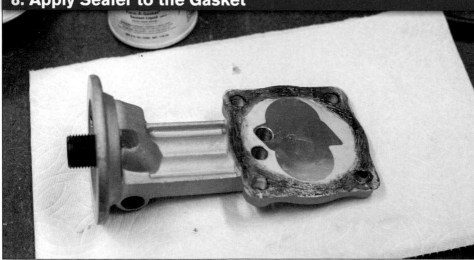

Apply another super thin coat of gasket sealer to the gasket supplied in the filter pump kit. Note the position of the inlet and outlet holes. DO NOT use the oil pump gasket supplied in the engine gasket kit.

9. Apply Sealer to the Screws

Apply sealer to the four socket head screws supplied in the filter pump kit. These are the correct length; the original pump studs are too short.

10. Install the Oil Pump Gears

Install the oil pump gears using plenty of assembly lube. Turn the driven gear until the tang lines up with the slot in the cam. Try to keep the sealing surfaces free of lube as much as possible.

11. Install the Filter Housing

Install the filter housing with the socket head screws supplied in the kit. Torque them to 14 ft-lbs using a 6-mm hex bit.

12. Install the Oil Filter

Install the oil filter supplied in the kit. Note that the filter housing can only be installed with the filter pointed to the 3-4 side of the engine.

13. Final Inspection

Double-check the filter housing clearance with a piece of paper folded over a few times. Any amount of clearance is all you're looking for.

Sump Plate

Air-cooled VW engines do not have an oil pan to speak of. The entire oil capacity (2.65 quarts) is located in the lower portion of the engine case. Our capacity will increase slightly with the addition of the filter pump. The total capacity will be slightly more than 3 quarts.

The factory oil sump plate serves two purposes. First, it is the means in which the oil is drained. It either contains the drain plug or in later models without an actual drain plug, the entire plate is loosened up to drain the oil. Second, it gives you access to the oil strainer. The oil strainer is basically a screen that the oil pickup sucks the oil through. It is by no means a filter. It stops larger items from getting drawn into the pump, but fine particles have no problem circulating through the entire oiling system.

The strainers are engine specific depending on the size of the tube in the oil pickup: 40-hp engines have 14-mm pickups. Single oil relief cases found in most single-port engines have 16-mm pickups. All dual oil relief cases in all 1,600 dual-port engines have 19-mm pickups. Cleaning this screen and being diligent about the frequency of oil changes are your only defense against wear in a stock engine. We purchased a new oil strainer for our engine.

1. Install the First Gasket

Install the first gasket from the gasket kit dry. Do not use sealer.

2. Install the Strainer

Install a new oil strainer. This may take a little pushing and shoving to get flush.

3. Install the Second Gasket

Install the second gasket. Again, install it dry with no sealer.

4. Set the Sump Plate

Set on the sump plate and six copper washers from the gasket kit.

5. Install the Acorn Nuts

Install the six special acorn nuts and torque to 5 ft-lbs. The nuts are very soft and strip out very easily. They are made to give way first before pulling the stud out of the case. Do not overtighten them.

6. Install the Drain Plug

Install the drain plug with a new sealing washer from the gasket kit. Torque it to 25 ft-lbs.

Valvetrain

The valvetrain in the top end of an air-cooled VW engine consists of eight aluminum pushrods and two rocker arm assemblies. The cam and lifters are a solid tappet design as opposed to a hydraulic tappet design. Solid lifters require periodic adjustment to keep the lash at the factory specification.

Hydraulic lifters are self-adjusting and don't need any maintenance. All the engines covered in this publication need periodic valve lash adjustments due to the fact that they are all solid lifter engines.

1. Lube the Pushrods

Apply assembly oil to both ends of the pushrods.

2. Install the Pushrods

With the engine turned 90 degrees in the stand, drop each pushrod down the tube. All of the pushrods are the same length.

3. Install the O-Rings

If the cylinder heads have recesses around the rocker assembly studs, install new O-ring seals from the gasket kit. If no recess is present, DO NOT install the O-ring around the stud.

4. Place the Rocker Assemblies

Place the rocker assembly on the studs and align the pushrod tips with the end of the rockers.

5. Install the Nuts

Install the nuts and wave washers with a 13-mm socket. Torque to 18 ft-lbs.

Rotate the engine in the stand 180 degrees and repeat steps.

Installing the Crank Pulley

1. Install the Pulley

Usually, the engine sheet metal that goes behind the crank pulley would go on now, but since this engine is destined for an off-road vehicle, we will leave it off. There is no seal at this end of the crankshaft.

2. Torque the Pulley

With a little assembly oil on the threads, torque the crank pulley bolt to 36 ft-lbs with a 30-mm socket. You may have to reinstall the flywheel lock to stop the engine from trying to turn over while tightening the bolt.

Adjusting the Valves

Once the pushrods and rocker assemblies are installed, the valve lash adjustment can be performed. It helps to have the crank pulley installed at this time but it isn't necessary. The crank pulley key is exactly 90 degrees before top dead center (TDC). So once the key in the crank is at 9 o'clock, either cylinder number-1 or cylinder number-3 will be at TDC. Which one can be determined by which one has both the intake and exhaust rockers loose.

The step-by-step photos explain the shortcut way to adjust the valves, but the universal way is also correct. Both the intake and exhaust valves need to be shut to adjust the correct amount of lash. Rotating the crankshaft to TDC of the compression stroke in each cylinder will achieve that. The firing order is 1-4-3-2. Rotate the crankshaft clockwise to adjust valves in order.

Cylinder number-2 and number-4 will have TDC 180 degrees from the other two cylinders, meaning TDC will be at 6 o'clock instead of 12 o'clock. If the pulley isn't installed, the key will be at 3 o'clock for TDC of cylinders number-2 and number-4. Either method is correct, and with practice this process will go rather quickly.

Don't waste a lot of time getting the valves adjusted perfectly on your fresh build. It just needs to be good enough to get the engine started. Once the engine has had a chance to warm up and cool down a few times, the break-in oil can be dumped and the valves adjusted more precisely. The factory recommended valve lash for all the engines covered in this publication is 0.006 cold (less than 122°F).

1. Find Top Dead Center

Rotate the crankshaft until the dimple on the outer edge of the pulley is in line with the split in the case halves. This dimple is always TDC on number-1 and number-3, no matter where the inner timing notch or notches are.

2. Set the Initial Lash

If both valves at number-1 are loose, then you are on the compression stroke. If they are not, then rotate the crankshaft 360 degrees and line up the dimple again. Starting at number-1 exhaust, adjust the valve lash so a 0.006-inch feeler gauge is a tight drag fit, meaning it can be pulled out and put back in without much trouble. The 0.007-inch feeler should not go in.

3. Set the Final Lash

Adjust the number-1 intake valve the same way. Double-check the lash after tightening the jam nut. Two more valves can be adjusted with the crankshaft in this position: the number-2 intake valve and the number-4 exhaust valve. This is due to the fact the lifter is still on the base circle of the cam at that point. Rotate the crankshaft 360 degrees and line up the dimple again. Number-3 should be at TDC. Adjust the lash on both number-3 valves, the number-4 intake, and the number-2 exhaust. That's it. We will readjust the valves after breaking in the engine on the stand.

Installing the Valve Covers

1. Apply Sealer

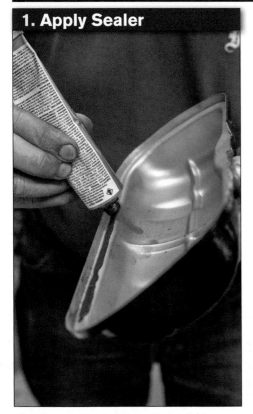

Make sure the inside of the valve cover is oil free with some brake cleaner. Run a thin bead of silicone sealer around where the gasket will seat.

2. Install the Gasket

Press the gasket into the valve cover.

3. Lube the Gasket

Wipe a thin coat of white grease or petroleum jelly on the gasket where it will seat on the head. This will prevent the gasket from sticking to the head when pulling the valve covers for future valve lash adjustments.

4. Lube the Valve Cover Bails

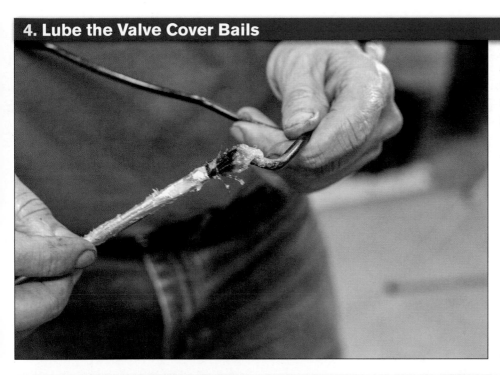

Add a little grease to the pivot ends of the valve cover bails and install them into the holes in the sides of the heads.

5. Install the Valve Cover

Install the valve cover with a little grease where the bail hits the outer shell of the cover. Just pry the bail until it snaps on. Don't be tempted to try the reusable rubber valve cover gaskets. They are a bad choice because they tend to squish out. There are no bolt holes keeping them in place.

Here is our completed long-block from valve cover to valve cover, flywheel to pulley. We are ready to bolt on the accessories.

ACCESSORY INSTALLATION

At this point, the engine was removed from the engine stand. The engine test run unit was installed on the engine and the engine was reinstalled on the same engine stand. A breakdown of how to build this test run engine stand adapter is covered in chapter 9.

Up to this point in the build, everything has been relatively simple and universal as far as putting a long-block together. Whether you have a 40-hp, a single-port, or a 1,600 dual-port, they all have the same amount of parts and go together the same way in the same order as far as a long-block goes. Now comes the fun part.

Our engine is destined to be installed in a 1970 Beetle that has been converted into a Baja Bug. The car won't have any heat and will have an off-road-style exhaust header and muffler. The cooling tin is simplified as well: having just a fan shroud and cylinder tin.

The oil pump we chose is the built-in oil filter style. This style oil pump will not fit with a factory peashooter-style muffler. When we had to decide what style oil pump we should use, we knew we weren't going to use a stock muffler, so the oil pump options were wide open.

Important Questions to Ask Yourself

A multitude of decisions have to be made and one decision will affect other options down the road.

What style exhaust are you going to use?

This should be the first thing you decide. If you're planning on using the factory-style muffler and the stock heater boxes, then everything is going to be stock and pretty straightforward. A stock muffler won't clear a filter oil pump. Everything else will remain stock as well.

With heater boxes, you will need to run the factory fan shroud or an aftermarket fan shroud with heat outlets. You can run a multitude of aftermarket mufflers from four-into-one-style headers with a separate muffler to all-in-one-style mufflers with chrome tips. It's all personal preference at this point, and none of these are super expensive to change in the future if you care to.

Are you going to have heat in the car?

As funny a question as this may seem, it's a very important one to ask yourself. Air-cooled Volkswagens weren't known for their stellar heat and defroster capabilities. Many owners choose to delete the heat for a variety of reasons. If you live in a warmer climate or if you are going to enjoy your VW only when the weather is favorable, then getting rid of the heater components is an option for you.

Heater boxes are heavy and costly. If you are performance minded, heater boxes are quite restrictive as far as exhaust flow is concerned. A nice, equal-length header exhaust system can free up a couple extra horses. An aftermarket fan shroud without heat outlets cleans up the engine compartment and directs 100 percent of the cooling air where it's needed most: cooling the engine. All of these factors may sway you one way or another.

Should I run the factory thermostat?

Wait. What? An air-cooled engine has a thermostat? Yes, remember when you removed a small rod connecting a set of flaps in the bottom of the fan shroud to a small brass accordion-looking thing attached to the side of the case? Well, that accordion-looking thing is the thermostat. It relies on warm air moving past it to warm it up and make it expand. It expands, that pushes the rod up, which opens the flaps and cools the engine.

The thermostat system serves two purposes. First, it allows the engine to warm up to operating temperature faster in cold weather, much like a water-cooled engine does. Second, by blocking air from cooling the engine it forces that air out the heater outlets and into the heater boxes and eventually into the car, warming up the cabin faster in cold weather. This all works great in a perfectly stock engine with all the factory-installed components. If any one item is changed or omitted in this system, it won't operate correctly, your thermostat won't open, and premature engine failure could occur.

The thermostat needs to live in this sealed, box-shaped area in order to get hot enough, quick enough. This box consists of the factory, peashooter-style muffler at the rear, the factory-style heater boxes at the sides, and the factory heat channel sheet metal that spans between the case and the heater boxes at the front and bottom. These three components must be in place for the thermostat to see enough hot air quick enough to expand.

If you plan on using a nonstock, performance exhaust or you aren't going to run heater boxes, you'd be wise to eliminate the thermostat and all of its components. It won't see enough heat to expand. If you aren't running heater boxes, then warming up the engine quicker or forcing more air through the heater boxes and into the car really isn't an issue. The engine will come up to operating temperature with or without the thermostat and its components, it'll just take longer in sub-freezing temperatures.

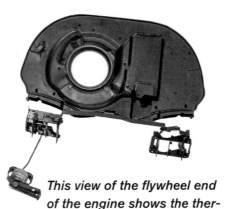

This view of the flywheel end of the engine shows the thermostat and its components. The two sets of flaps mount to the bottom of the fan shroud. The rod connects the thermostat to the flaps. The thermostat bracket is mounted to the case under the 1-2 cylinders.

Should I convert my 6V car to 12V?

Unless you are going for a 100-percent period correct restoration, converting your pre-1967 VW to 12V at this point makes perfect sense. The engine components you need to change are things you are already putting a wrench to: the generator, the voltage regulator, the ignition coil, and the electric choke on the carburetor. In our case, we went with a generator to alternator kit. It's internally regulated, so you can ditch the 6-volt regulator as well.

Can I install a newer 12V engine in my 6V car?

This is a super common swap and can easily be done. This is usually done along with a 6-to-12-volt swap of the entire vehicle. The biggest issue is the size of the flywheel. Most 6V VWs have a smaller 180-mm clutch, and all 12V flywheels use the 200-mm clutch. The transmission in a 6V car can easily be modified to accept an engine with the larger 200-mm clutch and flywheel.

The starter teeth on the flywheel will hit a few areas inside the bellhousing. These areas of the transmission case can be removed with an angle grinder. The starter and starter bushing will need to be replaced as well.

EMPI sells an adapter bushing to use a 12V starter in a 6V transmission

All of the components necessary for the factory thermostat to operate correctly are shown here: the factory peashooter-style muffler, factory heater boxes, and the lower engine tins (sleds) left and right. These items seal off the area around the factory thermostat so it only sees hot air from the engine above and isn't influenced by air passing under the car.

(part number 4027). A standard 12V starter (Bosch SR15) can be used in a 6V transmission as long as the starter bushing was replaced with the adapter bushing.

A way to get around replacing the starter bushing is to acquire a self-supporting starter more commonly known as an "auto-stick" starter. Bosch SR17X is the model for these heavy-duty starters that came in 1968–1972 Beetles with the clutchless automatic transmission called an "Auto-Stick." Once very pricey, they are more economical as of late and can save the hassle of changing out the starter bushing. We will be going over how to fit a 12V flywheel into a 6V transmission case in the next chapter.

Should I convert my non-doghouse oil cooler and shroud to a doghouse-style oil cooler and shroud?

With the increase of horsepower that came with the dual-port cylinder head design came an increase in oil operating temperature. To combat this, VW engineers moved the oil cooler element out of the fan shroud and into its own housing behind the fan shroud. They increased the size of the element itself and changed the element material to aluminum for better heat dissipation. The air leaving the oil cooler now exits outside the engine compartment instead of being forced on the 3-4 side of the engine cooling fins.

An increase in fan width from 32 mm to 37 mm not only decreased oil temperatures but cylinder temps as well. The decision to convert to a doghouse-style oil cooler should be made just as the original VW engineers came to theirs. Did you increase the horsepower or load put upon your engine? Is the strain on your vehicle going to increase such as driving at high speeds for extended periods? Will it be loaded down with extra weight such as a Type 2 bus might be? Do you live in Death Valley and this is your only means of transportation? All good reasons to convert.

For the most part, the cooling system of each engine was designed with a more-than-adequate capacity for the task at hand. However, if you are replacing some or most of the cooling items anyways, it might be in your best interest to consider upgrading your cooling system to the more efficient doghouse style.

Carburetor: Rebuild or Replace?

In 1970, a 1,600-cc single-port VW engine came equipped with a Solex 30PICT carburetor. It is very simple in design and very easy to adjust. In 1971, the Solex 34PICT found on all 1,600-cc dual-port VW engines was a fickle unit. The 34PICT has an internal idle circuit where the 30PICT uses the throttle plate to adjust the idle.

With the tiny passages of an internal idle circuit came a slew of problems. These passages would clog up, and once clogged they were nearly impossible to clear out. The introduction of reformulated fuels and the fact that most VW owners only drive their cars occasionally doesn't help the issue. Many older carburetors are beyond the point of being rebuildable.

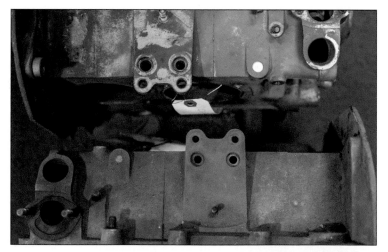

To mount a doghouse-style oil cooler to an early engine case (lower), two things must be modified. First, the two mounting holes in the case must be drilled out to 5/16-inch. Second, the mounting stud in the top of the case must be increased from 6 mm to 8 mm by drilling and tapping with a 8x1.25-mm tap.

To convert to a doghouse-style oil cooler, the two items on the right are replaced by the four items on the left. Not shown is the special doghouse firewall tin and the special thermostat flaps and linkage used with the later fan shroud.

The ones that can be rebuilt cost a pretty penny to get the job done correctly.

If your original carburetor works as is, don't disturb a good thing. If your engine idles and has good power with your original carburetor, consider yourself lucky. Replacement carburetors are readily available and usually run great straight out of the box.

Our engine came with a carburetor and rather than try our luck to see if it worked, we decided to opt for a brand-new one. EMPI 98-1289 is the part number for a replacement 34PICT carburetor. It comes nicely packaged with very detailed instructions. Upon opening the box, you may smell a hint of gasoline because all carburetors from EMPI are tested and preset on a test engine. Once you install the carburetor, only minimal adjustments should be necessary.

Installing Accessories

The order in which you install the accessories on the long-block is important. Some items need to be

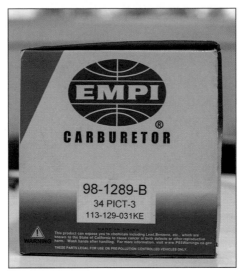

Rather than taking our chances rebuilding a more than 50-year-old carburetor, we chose to go with a brand-new unit. This 34PICT-3 replacement carb from EMPI is a good choice. They come nicely packaged and ready to run on a stock engine straight out of the box.

installed first, as other items will be in the way for them to go on later. If you follow the order in this book, you shouldn't have a problem.

The distributor drive was put off until this chapter on purpose. Until either the fuel pump or the distributor is installed, the only thing holding the distributor drive in place is gravity. Once the drive is installed,

it is imperative that one of these two items go on next. The fuel pump pushrod will hold the drive in place as it rides on the ramp built into the side of the drive. Without something holding the drive in place, it's possible for you to rotate the crankshaft the wrong way, forcing the drive up and out of the teeth on the brass drive gear and ruining it.

Distributor Drive

1. Inspect the Parts

The components that comprise the distributor drive assembly are the drive itself and the two thrust shims that set the drive depth. These items rarely need replacing.

2. Lube the Shims

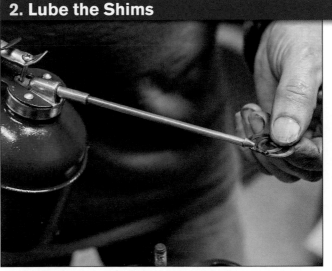

Verify that the crankshaft is rotated to TDC for the number-1 cylinder by lining up the outer dimple in the pulley with the split in the case and making sure both valves at number-1 are closed. Apply some assembly lube to the two shims that go under the distributor drive.

3. Install the Shims

Slide the shims over a long screwdriver. Place the end of the screwdriver all the way at the bottom of the distributor drive bore. Drop the shims down the hole and verify with a flashlight that they are centered. The oil should keep them in place.

4. Install the Drive

Lube the distributor drive and bore with assembly oil. With the snap ring pliers, install the drive with the slot toward the pulley. Compensate slightly for the angle of the teeth and twist that way while sliding the drive in. Note that the dimple on the pulley is still in line with the split in the case.

5. Inspect the Drive

This is how the distributor drive should look if installed correctly. The offset slot should be toward the pulley. Don't worry if it's not perfectly parallel with the pulley. It almost never is.

6. Measure Installation

With the distributor clamp installed on the distributor, measure 1.780 inches from the bottom of the clamp. It should land somewhere in the center of the two drive tangs. Some aftermarket distributors have been known to be shy in that dimension, and an extra shim must be added underneath the drive to cure that.

Here we are measuring the depth of the installed distributor drive. If it's installed correctly, it should measure approximately 1.780 inches. If it's short, then the drive isn't sitting all the way down against the shims. If it's long, you probably forgot to install the two shims before dropping in the drive.

7. Install the Spring

Another often overlooked item is this spring that goes between the end of the distributor and the drive. It keeps constant downward pressure on the drive to keep it meshing at the same point on the gear teeth. Without it, the timing can drift around as the drive moves up and down in the bore.

Fuel Pump Variations

Shown here are the fuel pump components: pump, correct length pushrod, base, and corresponding gaskets.

Mechanical fuel pumps come in a variety of styles but all serve the same purpose. Shown here are a rebuildable factory style (upper right), a typical aftermarket (lower right), a mushroom style for generators (lower left), and a 15-degree Mushroom style for alternators (upper left). All four are made in Brazil.

Alternator-equipped engines require an alternator specific fuel pump (left). It's basically the standard fuel pump (right) tilted at 15 degrees to clear the main body of the alternator. It also requires a shorter 100-mm fuel pump rod versus an 108-mm rod to operate correctly.

An easy way to tell the difference is the standard pump (right) has the paddle recessed below the mounting surface. The alternator style (left) has the paddle sticking past the mounting surface. The base under the pump is universal.

Install the Fuel Pump

1. Apply Gaskets, Sealer & Base

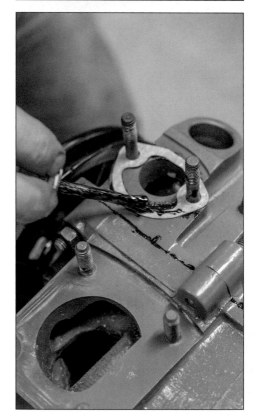

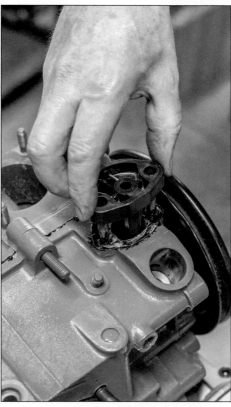

Once the distributor drive is installed, the fuel pump goes on next. Some gasket sealer, the base gasket, more sealer, and the fuel pump base are set over the fuel pump studs.

2. Oil the Pushrod

Assembly oil is applied to the correct fuel pump pushrod. We are installing an alternator-style fuel pump so the shorter 100-mm rod is used.

3. Install the Rod & Apply Sealer

Install the rod with the pointed end down, apply gasket sealer to the base, place the other fuel pump gasket over the studs and rod, and apply more sealer to the gasket.

4. Grease the Pump

Apply a small amount of grease to the lever at the bottom of the fuel pump. This item doesn't see much lubrication during its service, so this is vital to its survival.

5. Install the Pump

Install the fuel pump with a 13-mm wrench. Don't forget the wave washers under the nuts. If the fuel pump seems too high off the base, don't force it on with the nuts. This indicates that two things may be wrong: either the distributor drive isn't seated all the way down in the bore or the fuel pump pushrod is the wrong length.

Alternator Stand

The generator/alternator stand is easiest to install before the intake manifold. You can get at all four nuts without anything in the way except the fuel pump, which isn't much of an issue.

1. Apply Sealer & the Baffle

The next item to install is the generator/alternator stand. Apply gasket sealer to the engine case and install the engine baffle/base gasket. Make sure it is installed as shown. It can be installed four different ways, but only one way is correct.

2. Apply Sealer & the Alternator Stand

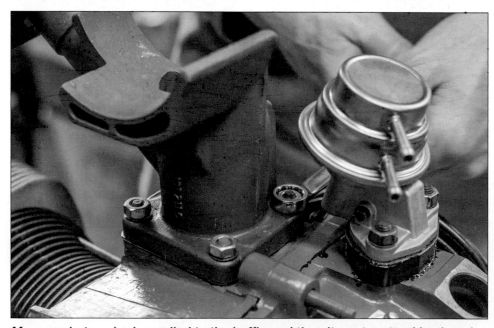

More gasket sealer is applied to the baffle and the alternator stand is placed over the studs. Install four nuts and wave washers and tighten with a 13-mm wrench.

Oil Cooler

The oil cooler is much easier to install before the 3-4 side cylinder tin. The two nuts holding the oil cooler or oil cooler adapter to the case are visible, but you can't get your hand in there to start the nuts. Leave the cylinder tin off until the oil cooler is installed.

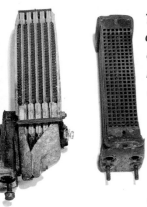

The doghouse-style oil cooler (left) needs to be assembled to the adapter before bolting it to the case. The non-doghouse cooler (right) used on all single-port and 40-hp engines bolts directly to the case. The seals are different as well. The doghouse style has flanged seals. The non-doghouse seals are straight. Both styles should be in the gasket kit.

All of the components to a doghouse-style oil cooler are shown: top is the oil cooler and under that is the cooler adapter, seals, and the hardware. Lower left is the oil cooler sealing tin, more commonly known as the "hoover bit."

Install the hoover bit over the two longer studs as shown. Install the three nuts and lock washers. Tighten these with a 10-mm wrench. Don't get carried away tightening these nuts. Self-locking nuts here are also a good idea.

The green oil cooler seals in the gasket kit go between the oil cooler itself and the adapter. Don't try to use any other seals because they are too thick and the adapter won't butt up flush with the oil cooler.

Place the orange seals in the oil passages for the oil cooler adapter. These are the correct seals for our application. The other seals in the gasket kit are for a non-doghouse-style oil cooler such as a single-port or a 40-hp engine would have.

Place the oil cooler assembly on the case. Install the three nuts and wave washers with a 13-mm wrench. There are two nuts underneath and one is on top of the case. These have a habit of loosening up, so be sure these are good and snug.

Cylinder Tin

Here are examples of the three main cylinder engine tin sets: single-port, all years (top); 1,600-cc dual-port (middle); and factory fuel-injected Beetle 1975–1979 (bottom). Not shown is the 40-hp cylinder tin, which is similar to the single-port tin. Both single-port and dual-port cylinder tins are available through the aftermarket.

The top of this photo shows the nice deflector inside the factory cylinder tin to redirect the air evenly to each side of the cylinder head. Most aftermarket cylinder tins don't have this.

Next, the cylinder tin can be installed on both sides. These are held on by two 6-mm screws in each head. It's a good idea to add a little anti-seize to the threads of these screws. Don't overtighten them.

Intake Manifold

The intake manifold must be installed before the fan shroud is dropped over the oil cooler. Even though the intake manifold is in three pieces, the right side heat riser on the center manifold section won't fit between the generator stand and the fan shroud unless you take the stand off.

The entire intake manifold for a dual-port engine is shown here. The intake gaskets are from the gasket kit. Left and right end castings have been blasted and painted. We are using new manifold boots (EMPI 3404) and the original clamps. The center manifold section has also been blasted and painted.

The 40-hp and single-port engines do not use standard gaskets to seal the intake manifold to the cylinder heads. Thin crush rings that are supplied in the gasket kit do the job. These are a one-time-use item, and make sure you use the right ones out of the kit. The 40-hp engines use the smaller ones (right).

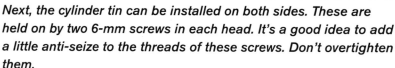

1. Place Gaskets

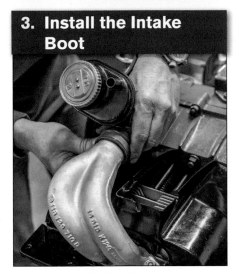

Apply gasket sealer to both sides of the metal intake gasket from the gasket kit. Place them over the studs on each head. Note the extra-small hole in the gasket between the ports. Some heads have a dowel to locate the manifold on the head. Very few heads have this dowel.

2. Install 1-2 End Casting

Install the 1-2 side intake end casting with two nuts and wave washers. Get the hardware started and don't tighten them yet. Once all the components are installed, we will tighten them in order.

3. Install the Intake Boot

Loosely install the clamps on an intake boot and slide it on the intake end casting. A little oil on the inside of the boot will help it slide on. Tighten the larger clamp but leave the small one loose.

4. Add Washers

The one case half stud that's longer than the rest is the one that mounts the center manifold section. A thick washer and a wave washer and nut go here.

Again, don't tighten them, just leave them hand tight for now.

7. Tighten All

Once the two nuts and wave washers are installed on the 3-4 side intake end casting, we can start tightening everything up. Tighten both end castings to the heads first. Then tighten the one nut that holds the center section to the case. Finally, snug up the four clamps on the intake boots.

5. Install the Center Intake Manifold

Push the center intake manifold into the boot. Again, a little oil inside the boot will help it slide. Some twisting will be necessary to get it to go in enough.

6. Install 3-4 Boot

Install the other boot with loose clamps on the intake end casting for the 3-4 side of the engine. Twist it onto the center section. By keeping all the hardware loose, it's easier to align everything.

Fan Shroud

We are using an aftermarket fan shroud on our build. More commonly known as a 36-hp fan shroud because its shape is based on the original VW fan shroud made for early 36-hp engines. Aftermarket 36-hp fan shrouds come in four varieties: doghouse or non-doghouse oil coolers and heat or no heat.

Doghouse would be all dual-port engines and non-doghouse would be 40-hp and single-port engines. Fan shrouds with ducts that connect to the heater boxes can be purchased in either doghouse or non-doghouse style. The fan shroud used in our build is a doghouse style without heat (EMPI part number 8672).

All of these fan shrouds are produced without threaded holes to mount the coil. They also are missing the holes to install the thermostat flaps and the spark plug wire holders. They do have the threaded holes for the generator/alternator tin, the cylinder tin, and the oil cooler tin.

On our build, we mounted the coil on the upper portion of the fan shroud for a number of reasons. VW engineers mounted the coil upside down on the face of the shroud.

Most coil manufacturers recommend mounting an oil-filled coil with the terminals pointing up. Ever notice how the coils are mounted on a buggy racing in the Baja 1000? In case of a leak in the coil case, most of the oil will remain inside and keep the coil cool and functioning. We compromised and mounted it horizontally. It gives the engine a cleaner look and you can identify the terminal marking much easier.

In the case of an engine with aftermarket dual carburetors, the factory coil location is in the way of most dual carb throttle linkages. Mounting it horizontally alleviates that problem. The only drawback of mounting the coil up higher is that the primary coil wire will have to be longer.

The doghouse fan shroud uses an adapter to locate the oil cooler outside the direct path of the cooling air stream. This particular shroud is from a fuel-injected model, which is determined by the velocity ring around the air inlet.

The difference between a factory non-doghouse and doghouse fan shroud is shown here. The left shroud is off a 1,600 dual-port, and the oil cooler resides outside the main housing. The right shroud is from a 40-hp or single-port engine, and the oil cooler resides in the main housing.

The aftermarket 36-hp-style doghouse fan shroud (left) is available with or without heat outlets. The factory doghouse fan shroud (right) shows its typical 50-plus years of dents and extra holes. A new fan shroud will make a world of difference on how the finished product looks.

A typical 40-hp or single-port engine fan shroud and oil cooler is shown. The oil cooler resides in the main cavity of the shroud, directly in the path of the cooling air going to the 3-4 cylinders. This is not an ideal situation.

Mount Coil to Fan Shroud

1. Mark Spot on the Shroud

Aftermarket fan shrouds don't come with any provisions to mount the ignition coil. Place a straightedge in line with the upper generator mounting holes. Mark a spot on the edge of the fan shroud in line with the straightedge and midway on the edge.

2. Drill the Shroud

Drill a 3/8-inch hole at the mark you made with the straightedge. A step drill works great in sheet metal, just make sure you stop at the correct step.

3. Use the Threaded Inserts

Threaded inserts are great for installing threads into sheet metal. These are 1/4-20 threads and require a 3/8-inch hole. You can find them at any hardware store, and the kit comes with the installation tool.

4. Install the Insert

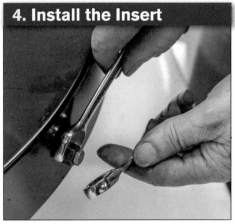

Screw an insert into the insertion tool. Place the assembly in the 3/8-inch hole. Hold the base of the tool with one wrench and tighten the bolt with the other wrench. The insert will expand and grip the sheet metal.

5. Use the Coil Bracket as a Template

Temporarily install the coil and mark where the upper threaded insert will go to hold the coil. Drill another 3/8-inch hole there and install another threaded insert.

Alternators and Generators

Shown here is a typical 12V generator and fan assembly. This entire unit is held to the fan shroud with four 6-mm bolts. The generator can be tested, rebuilt, or replaced with a new one. With 6V vehicles, new generators are very hard to find; rebuilding your original is your only choice.

Here is a 12V generator and fan assembly blown apart, cleaned up, and ready for reassembly.

Three different versions of 12-volt charging units are: the factory generator (left), the factory alternator with external regulator (center), and an aftermarket 55-amp internally regulated alternator (right).

Your typical generator setup had an externally regulated generator, generator-only stand, standard fuel pump, and a 108-mm fuel pump rod. This is a 12V example but the 6V setup will be the same. The fuel pumps are the same for all generator-equipped vehicles whether 12V or 6V.

Convert from Generator to Alternator

1. Inspect Parts

All of the components necessary to convert a generator to an alternator on any Type 1–based engine are a 55-amp internally regulated alternator, an alternator stand, a 15-degree fuel pump, and a 100-mm fuel pump pushrod.

2. Remove the Tape

A new alternator comes with new keys on both ends of the shaft: one for the pulley and the other for the cooling fan hub. Remove the tape that keeps them from falling out.

3. Install the Pulley

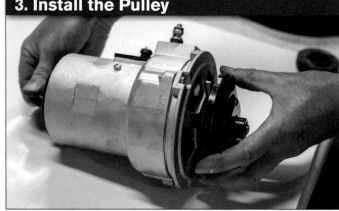

Align the keyway in the back half of the pulley and push it on. Add a little lube to make this easier. It'll also stop the two from corroding together if you ever need to replace either of them.

4. Add Shims

Add the adjusting shims temporarily. You should have at least eight shims. These shims will be used to adjust the tension on the belt by adding or subtracting them from between the two pulley halves.

5. Add the Outer Pulley Half

Add the outer pulley half by aligning the two tabs into the two slots in the back side pulley half. Note that the slots are different widths and correspond to the two different width tabs.

6. Install the Spacer & Nut

Install the outer spacer and nut temporarily to keep the pulley end components all together. This gives us something to set the alternator upright on, enabling us to put together the fan end of the assembly. Once we get the fan end together, we can use the pulley to spin the shaft and check for clearance issues.

7. Install the Fan Cover

With the pulley end sitting on some blocks, we can begin the fan end assembly. First make sure the outer fan tin has the correct side out. If it has a small offset hole, this hole should be behind the stand when installed in the vehicle.

9. Place the Inner Tin

The inner alternator tin has a slot in it. This slot must face down when installed in the vehicle. An easy way to know it is in the right way is when the slot is opposite the electrical terminals when installed.

8. Install the Reinforcement Ring

Next is the outer tin reinforcement ring. This ring sandwiches the tin to the backside of the alternator and creates a seal for the air passing through the body of the alternator for cooling it.

10. Tighten the Fan Cover Assembly

The inner tin is mounted with two small washers and two 6-mm nuts. Self-locking nuts are a good idea at these locations. The last thing you need is these coming loose and getting trapped behind the fan. Tighten them up with a 10-mm socket.

11. Install the Flange

Next install the fan drive flange. New alternators come with the key installed. Line up the keyway in the flange with the key. Add some lube, as we did on the pulley end. It should just push on.

12. Install the Shims

Two shims go between the drive flange and the fan itself. These are the same shims that are used to adjust the belt tension at the pulley end. These shims should create the correct 0.060-inch gap between the back side of the fan and the inner tin.

13. Verify the Fan Width

The difference in the two fan widths is shown here. The left is a 40-hp or single-port cooling fan and it measures 1^1/$_4$-inch wide. The fan on the right is for a 1,600 dual-port and measures 1^7/$_{16}$-inches wide. Narrow fans fit in fan shrouds with the oil cooler directly in the shroud. The wider fans fit the doghouse-style shrouds.

16. Install the Fan Nut

Install the special fan nut and torque it to 43 ft-lbs using a 36-mm (1^7/$_{16}$-inch) socket. The same size socket as the flywheel gland nut and the axle nuts that hold the rear brake drums on. It's a handy size to have.

14. Align the Flats

Align the flats on the drive flange with the flats on the fan opening. If the flats in the fan are not completely straight and the opening looks more like a bow tie, then the fan mounting nut came loose at some time and beat that opening up. Time to find a new fan.

15. Install the Wave Washer

The special wave washer goes on next. It has flats like the fan does but it also has a dish to keep it tight. Make sure the sides of the washer opposite the flats are facing up and out. The recess in the nut goes toward the washer.

Our new alternator is assembled using our old generator tin, pulley, and fan and is ready to be installed into our fan shroud. It's sitting on a factory alternator stand that is ready to be mounted on our long-block. Due to the shape of the alternator, it won't fit on a factory generator stand.

Mount the Alternator in the Fan Shroud

The easiest way to install the generator/alternator and fan assembly is to install them into the fan shroud ahead of time and drop the entire fan shroud/cooling fan/charging system down on the engine all at once. They can be installed separately, but it's much easier to install as an assembly.

Drop the generator/alternator assembly into the fan shroud. Make sure the vent slot in the backing plate is facing down and the B+ terminal is facing up when on the engine.

Fasten the assembly to the fan shroud with four short 6-mm bolts. Use a 1/4-inch drive ratchet and a 10-mm socket to limit the torque on these fasteners. They just need to be snugged up.

Wrap the mounting strap around the alternator and loosely install the hardware. It's much easier to do this now than to try and fish it under the alternator stand and get the nut and bolt started.

Install the Fan Shroud on the Engine

Install two short 6-mm bolts with larger washers on either end of the fan shroud. Install one short 6-mm bolt into the hoover bit at the oil cooler. Carefully place the fan shroud assembly over the oil cooler and make sure the generator clamp isn't pinched under the generator stand. Guide the thermostat rod down between the cylinders if you are using a thermostat.

It may take a bit of finagling to get everything in place. The cylinder tin needs to be on the outside of the shroud. The bolts and washers need to be on the outside and in their slots. A small indentation underneath the alternator needs to fit into a protrusion in the alternator stand.

Once the fan shroud assembly is situated, tighten down the generator clamp with a 13-mm socket and a 13-mm wrench. Now, tighten the three short 6-mm bolts with a 10-mm wrench. Give the alternator a spin to make sure the fan isn't rubbing on anything. The thermostat (if you are using one) can be threaded onto the rod and secured to the bracket under the 1-2 side pushrod tubes.

Throttle Cable Tube

Install the throttle cable tube once the fan shroud is mounted. It is much harder to fish through the shroud with the distributor and carburetor installed. It is nearly impossible to get the throttle cable to the carburetor without this tube, so don't forget to install it. The thermostat flap linkage behind the fan shroud can be installed now if so equipped. Attach its return spring as well.

All the sheet metal behind the fan shroud can also be installed at this time. On doghouse-style fan shrouds, install the firewall tin, then the oil cooler chute, and then the shroud-to-chute connector tin. On non-doghouse fan shrouds, the firewall tin can be installed.

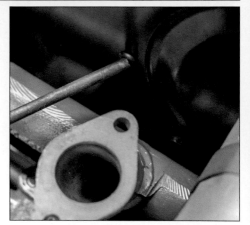

Once the fan shroud is secured, now is a good time to install the throttle cable tube. This hole is predrilled from the manufacturer through the entire shroud. Place a rubber grommet in the hole to hold the tube in place during operation.

Firewall Tin

When installing an engine in a stock Beetle, all of the engine tin needs to be installed behind the fan shroud. The hot air from the oil cooler gets routed downward with a two-sided piece and a chute that goes through the rear firewall tin. Non-doghouse fan shrouds just have the firewall tin without the hole for the chute.

Pulley Tin

The pulley tin mounts to the engine case before the pulley is installed if the engine is being fitted into a stock Beetle. The rear apron tin goes on after the engine is installed in the car. This one has holes for the heater tubes to connect the fan shroud to the heater boxes and the air cleaner to the preheat tube.

Dipstick

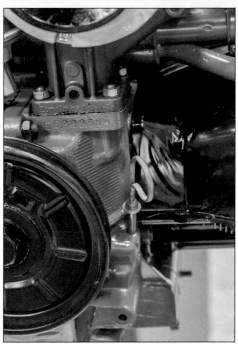

All the direct-fit dipsticks are all same length; only the size of the finger loop changed. The loop got bigger as the engines got larger. A Type 3 engine won't have a dipstick directly to the engine case. The dipstick and its tube are mounted directly to the body of the car.

Lower Cylinder Tin

Many times these two items are missing from the cooling tin components, but they serve an important purpose in the grand scheme of things. They redirect the cooling air around the sides of the cylinders, eliminating hot spots.

Two rather important pieces of cooling tin are often overlooked. These small items mount to the front of the cylinder tin and redirect the airflow around the sides of the barrels. Omitting these items can cause a hot spot in the barrel and warp the bore.

Oil Pressure Sender

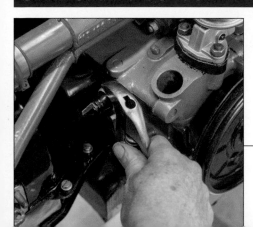

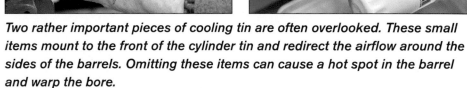

The oil pressure sender is often abused when it is being installed. The key here is to tighten it just tight enough so it doesn't leak. It has tapered threads, and it is very easy to overtighten and ruin the case. Do not put Teflon tape on this because it needs to ground out the oil light circuit.

Carburetor Preparation

1. Inspect Parts

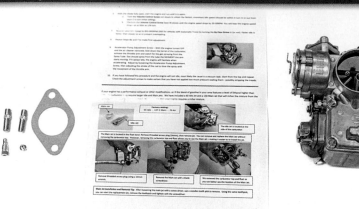

Here is everything you get with the carburetor: mounting studs, base gasket, larger 130 main jet, larger 60 idle jet, and very detailed instructions on how to initially set up the carburetor and adjust the idle and mixture.

2. Upgrade Main Jets

Since we already know our engine will have a better flowing exhaust than stock and our general area has fuel that's at least 10-percent ethanol, we chose to change out the stock 127.5 main jet for the larger 130 main jet. It's easier to do it now before the carburetor is mounted to the manifold.

3. Upgrade Idle Jet

The idle jet is on the electric choke/accelerator pump side of the carburetor. We are swapping out the factory 0.55-mm jet for a larger 0.60-mm jet included with the carburetor. This, along with the larger main jet, will help our engine achieve maximum efficiency due to the fact it has a better flowing exhaust.

4. Install Mounting Studs

The only assembly necessary before installing the carburetor on the manifold is to screw in the mounting studs. A small amount of red Loctite is put on the short end of the stud.

Snug up the studs with needle-nose pliers on the knurled part of the stud. No need to double nut them to drive them in. The red Loctite will do its job.

Electronic Ignition

All of the carbureted engines covered in this manual have standard points–type ignitions. As with many electronic devices, technology has improved them drastically. They are more efficient and less expensive than years past. Why not replace the old electronics with something that should last indefinitely, never need adjustment, and provide more efficiency.

We will replace our points and condenser with an electronic ignition or basically a points eliminator. Once we set the ignition timing, that's the last time we should ever have to attend to anything underneath the cap and rotor. The ignition dwell will always be spot on (44 to 50 degrees). Adjusting the point gap to achieve that is a thing of the past. The timing will stay rock solid through the entire RPM range.

The price for the electronic ignition kit is $40 and the price for replacement points and a new condenser is $21. You get a return on your investment the next time they need to be replaced.

This is the electronic ignition kit we will be installing in our distributor. The EMPI kit (part number 9432) fits all distributors that use Bosch 044 points, as it is basically replacing them. Note this kit must be used with a coil with an internal resistor or an external resistor must be added.

4. Install Mounting Studs
continued

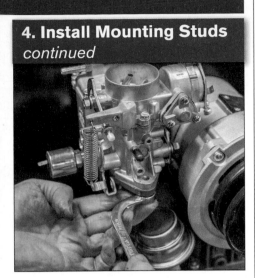

A 13-mm S-shaped wrench makes getting at the back nut very easy. This carburetor is designed to clear an alternator. In some cases, a riser block between the carburetor and the manifold will be necessary to clear it. Both single-port and dual-port versions of this spacer are available.

5. Inspect Installation

A new base gasket is supplied with the carburetor. Notice all the vacuum ports are capped off. One on the manifold and two on the carburetor itself. That's because we are running a 100-percent mechanical advance distributor, and vacuum won't be necessary.

The components of our electronic ignition include the module, trigger disk, hardware, electrical terminals, 3-mm wrench, and instructions. The special spacer is included to fit multiple models.

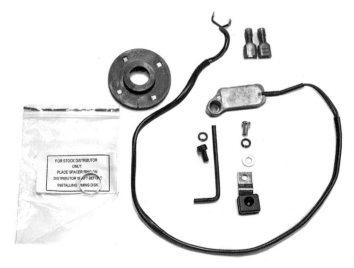

Electronic Ignition Installation

1. Inspect the Distributor

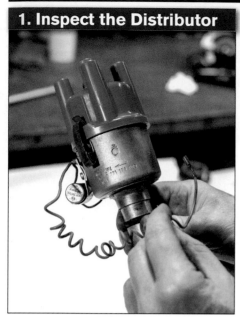

The distributor we are converting is the ever popular Bosch "009" (named so by the last three digits of the part number). This mechanical advance distributor is a direct replacement for vacuum advance units VW installed on almost every engine covered in this book.

2. Remove the Cap

Remove the cap by prying back the two clips. Pull the rotor off and remove the dust shield. The dust shield won't be reinstalled because the electronic unit is dust and water resistant.

3. Remove the Wire

Remove the wire from the points to the condenser. One small screw holds the points in place. This will be the same threaded hole that holds the module in place but with a small socket head cap screw and lock washer.

4. Remove the Condenser

One screw on the outside holds the condenser to the distributor body. The condenser bracket also holds the grommet and wire to points in place. This will all come out as one piece.

5. Remove the Points

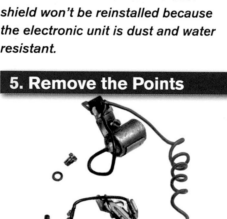

These are the parts we are eliminating: the points and condenser along with all their hardware. It's never a bad idea to place all these items in a small plastic bag and put the bag in the glove box. In the rare chance your electronic ignition fails, at least you have the parts to put you back on the road.

6. Install the Grommet

Install the rubber grommet from the kit into the square hole that the condenser wire passed through. Note that the grommet has one end shortened up. That end goes toward the cap, as the cap and that grommet want to occupy the same space.

7. Thread the Wires

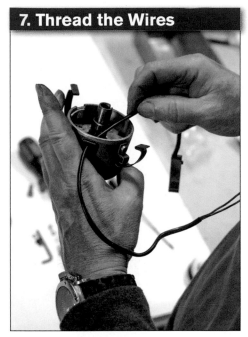

Fish the wires for the module through the grommet from the inside out. The terminals were left off the wires so this can be done.

8. Place the Module

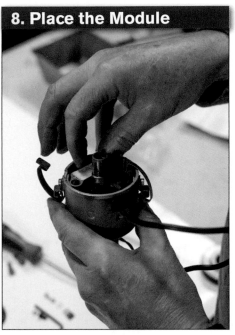

Place the module where the old points were. A small protrusion in the bottom of the module locates the unit in the point plate.

9. Secure the Module

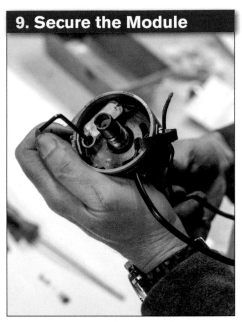

Secure the module using the small socket head cap screw and lock washer included in the kit. The 3-mm Allen wrench included in the kit is to tighten this one screw.

10. Check the Wire Routing

Make sure the wires inside the body are tucked away from the only moving part: the shaft. Secure the grommet with the tab and hardware included in the kit.

11. Install the Trigger Disk

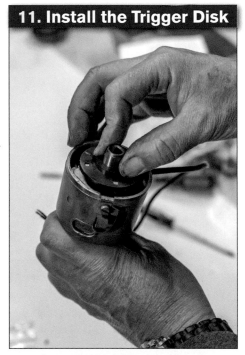

Install the trigger disk onto the shaft. The squarish recess in the bottom locks onto the lobes of the shaft to locate it. The magnets embedded in the disk will be just above the module if installed correctly.

12. Check the Spacer

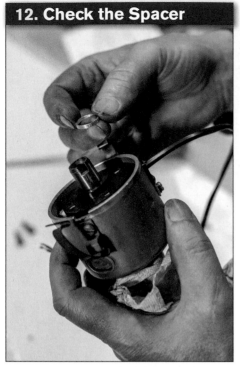

This small aluminum spacer is NOT to be installed on a 009 distributor. It is used on all vacuum advance distributors that use Bosch 044 points. It will be nearly impossible to install the cap with this spacer installed but it's been done.

13. Install the Cap & Rotor

The cap and rotor can be reinstalled now. The rotor pushes on as before and will keep the trigger disk down. The tab in the cap lines up with the slot in the distributor body by the clip.

14. Finish Wiring

Last thing to do is crimp the terminals supplied in the kit on the ends of the wires. This entire process took less than 20 minutes. Once you set your timing, you can "set it and forget it."

Distributor

With the seal lubed up and the tangs positioned at roughly 4 and 8 o'clock, install the distributor by lining up the tangs with the slot in the drive. Rotate the rotor while pushing down. Get the clamp lined up with the stud in the case as well.

Connect the electronic ignition to the coil. There are only two wires, but if you put the wrong wire on the wrong terminal you will ruin the module and usually there is no return on electrical parts. Positive (RED) goes to the terminal marked "15"; negative (BLACK) goes to the terminal marked "1." To help you remember, just think of the "1" terminal as a minus (—) sign.

Though hard to see in this photo, there's a small notch on the edge of the distributor where the cap sits. That notch is number 1 in the firing order. If everything is installed correctly, the rotor should be at the 4 to 5 o'clock position. Rotate the distributor body to line up the rotor with the notch. Remember, this is all with the engine sitting at number-1 TDC.

Install the one nut and wave washer that holds the distributor clamp down with a 13-mm socket and torque it down to 14 ft-lbs.

Install the ignition coil and its clamp to the fan shroud using 1/4-20 bolts and lock washers. Remember, the threaded inserts we installed earlier are only aluminum. Just snug them up.

Spark Plugs

The factory spark plug for all the engines covered in the book is a Bosch W8AC. Gap all the spark plugs to 0.028 inch. These handy gapping tools are much easier to use than feeler gauges.

Apply a small amount of anti-seize on the threads of the spark plug. A great way to get the plugs started is by slipping a short length of tubing over the porcelain of the plug. You get a feel for how straight you are installing the plug and the chance of cross threading them is greatly reduced.

Snug up the spark plugs with a 13/16-inch spark plug socket. The correct torque is 22 to 29 ft-lbs, but it's recommended to stick to the lower value.

Spark Plug Wires

The two shorter spark plug wires are for the 3-4 side of the engine and are easy to install. Number-3 is opposite the number-1 notch in the distributor housing and number-4 is 90-degrees counterclockwise from that. Both are on the left side of the distributor.

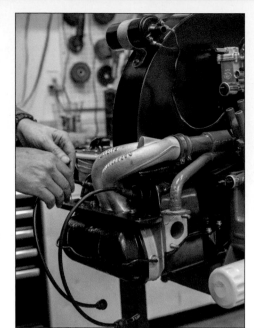

The two longer spark plug wires get fished under the alternator. Number-1 is the terminal above the notch in the distributor housing. Number-2 is 90 degrees counterclockwise from that.

Since our aftermarket fan shroud doesn't have provisions for the factory spark plug wire clips, we are going to make our own wire loom using zip ties. Gather the wires together with one zip tie and separate them with another zip tie.

Tie the spark plug wires together in two spots in each set. Snip the zip tie ends off flush and you have a nice, clean, inexpensive way to organize the spark plug wires.

Here is how your spark plug wires should look coming out of the cap. Number-1 should be at the lower right terminal, right above the notch in the distributor housing under the cap. From there it's rather simple. In a counterclockwise direction, it's 1-2-3-4.

Fan Belt

There is no idler pulley or way to move the alternator to adjust the tension on the fan belt. The tension is set by these 10 shims. The shims go between the two halves of the pulley to make it wider or narrower. Start with 8 of the 10 shims in between for your initial adjustment. Put the extra shims on the outer side of the pulley for future use.

The belt tension is adjusted by adding or subtracting shims between the two halves of the generator pulley. This is a trial-and-error adjustment, but once it's set, the adjustment should last some time.

The belt has enough tension when you can turn the engine over by tightening the generator nut without the belt slipping. If the belt slips, remove a shim from between the pulley halves and try again.

Once the tension on the belt has been established, snug up the pulley nut by inserting a large screwdriver into the slot in the backside of the pulley and rest the tip of the screwdriver against the alternator housing to stop the pulley from turning.

Fuel Line

Plumb the fuel pump (upper fitting) to the inlet on the carburetor. By using OEM-style 5-mm braided fuel line you can be sure of a snug fit on all the fittings. Hose clamps can be added for extra insurance.

Cut the fuel line midway between the pump and carburetor and add the fuel filter. The filter will have an arrow showing the flow and (as with most filters) the direction is from the outside of the element inward.

Engine Wiring

A couple of electrical connections on the carburetor need to see 12V when the ignition switch is turned on. Run a wire from the positive (+) side of the coil to the fuel shut off valve and the electric choke. Solid black wire was used as that is the color VW uses for all the 12V switched power.

Air Cleaner

An entire chapter could be written on factory VW air cleaners, but we will just scratch the surface. The top one is a typical oil bath–type air cleaner. This one is a 1967 model with two small preheat tubes. The center one is a factory plastic air cleaner that uses a panel-type pleated element, like most modern cars. It also has a vacuum-operated pre-heat device. The bottom one is a small chrome aftermarket air cleaner.

Add Oil

Sad as this may sound, many a fresh engine has been ruined from the simple mistake of not filling the engine with fresh oil before a test run. The engine oil pressure priming process was skipped and the bearings were ruined as a result.

Exhaust System

The exhaust we are installing on our fresh rebuild isn't the final exhaust that will be installed in the car. There's a couple reasons for this. First, we are going to test run our engine on the engine stand. A specific exhaust is used for this that is easy to install and has a remote muffler and tailpipe that can be directed away from the engine. We are going to start our engine in the shop and have the exhaust exit outside under the garage door. The exhaust exiting outside will help us hear the engine better and aid in diagnosing any strange noises we might encounter.

Second, we are using our test stand exhaust because the off-road exhaust that will eventually be installed covers up quite a lot of the front of the engine and would be in the way.

It would be difficult to cover all the many variations of exhaust combinations, so we will highlight the basic procedure only. The exhaust you install will go on in the same order just with more or fewer components.

Installing the Exhaust System

1. Place the Gaskets

The four metal exhaust gaskets found in the gasket kit can be placed on the four exhaust ports. A small amount of anti-seize is applied to the threads. Note the exhaust we are installing is only temporary and doesn't have provisions for the intake manifold preheat tubes.

2. Install Exhaust Pipes

The exhaust pipes from the number-1 and number-3 cylinders are installed first. In our case, they are heater box eliminator pipes called J-pipes. If you have heater boxes, they are installed now along with the lower sheet metal between the case and the heater boxes. Do not tighten the nuts up completely; leave them loose for now.

3. Add Tubular Exhaust

In our case, the tubular exhaust header is installed next. This would replace the factory muffler on a stock vehicle. Note how close the oil filter is to this header. A stock muffler would not fit at all.

4. Tighten the Flanges

With all the pipes fitted into their respective pieces, tighten up the flanges at the heads. The intake manifold preheat tubes are attached after that. Metal gaskets (similar to the exhaust gaskets but much smaller) are used between the manifold and the muffler.

5. Tighten Lower Connections

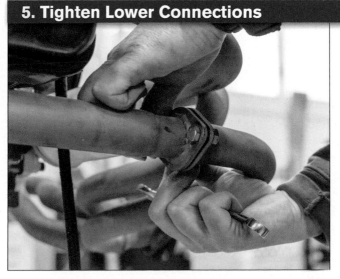

The factory lower exhaust connections between the heater box and muffler are usually an exhaust donut and two-piece clamp arrangement. In our case, that has been modified to a flange arrangement using a couple of exhaust flange kits (EMPI part number 3640).

STARTUP, BREAK-IN, AND ADJUSTMENTS

Congratulations! You have finished assembling your engine and are ready to see if all your hard work is going to pay off. Few things in life are as satisfying as putting an engine together and firing it up for the first time. All kinds of crazy thoughts race through your head. Did I torque this? Did I forget to install that? All those thoughts will be put to rest and you will have a sense of relief like no other.

One of the beautiful things about the VW air-cooled engine is that it's all self-contained. No separate cooling system to worry about. No radiator or coolant to mess with. The exhaust system is entirely attached to the engine. It only needs fuel and 6 or 12 volts and it'll run all on its own until it runs out of fuel. That's why it makes such a universal powerplant for so many vehicles and other types of equipment, such as water pumps, backup generators, weed cutters, log splitters, and sawmills.

You don't even have to worry about motor mounts because there

Rather than installing our fresh engine in the vehicle to fire up and break in, we are going to run it on a test stand. We will go over making this engine starter adapter for our engine stand. Testing it before we install it gives us a chance to do a more thorough job adjusting and checking for issues.

Basic Test Stand Items

- Heavy-duty universal engine stand: As mentioned earlier, they are inexpensive and easy to come by. The base for our run stand is from Harbor Freight (part number 47033) for $59.99.
- VW transaxle: Ideally, one that you won't mind cutting in half. It would also save some time if you obtained the same vintage transaxle as the engine you will be starting (6V or 12V). Junk 6V transaxles are easy to come by, and we will go over how to make that work with a 12V engine. Late-model IRS transaxles with only one axle side cover won't work, as we need mounting points on both sides. A late-model IRS VW bus transmission bellhousing makes a great engine starter base, but lately they have become rare and expensive. We are going the opposite route. You don't need the entire transaxle either, the only part that matters is the bellhousing and starter area. The rest can be junk or missing. Many shops that rebuild VW transmissions will have all kinds of transaxle cases laying around. I'm sure they would gladly let one go for scrap price or even less, possibly free. Ask around; many would gladly help on your project given the chance to borrow it once it's complete.
- Matching starter: 6V or 12V. If worse comes to worse, you can "borrow" the starter from the vehicle you are building the engine for.

are none. The engine hangs off the back of the transmission by four bolts. The transmission is mounted to the chassis and also houses the starter. By bolting up a transmission, in basically any condition, you can start an engine just about anywhere. You can start it directly on the floor if necessary. All you need is a way to start the engine, the proper battery, and either a 6V or a 12V starter.

Be sure you have help when removing the engine from the engine stand. Each individual part was rather light, but all of them assembled is a different story. Depending on how complete the engine is, it'll weigh between 200 and 250 pounds. Two adults of average strength should have no problem lifting it. Remember to lift with your legs, not with your back. Once down on the ground, the engine stand yoke can be removed. At this point, it can be installed in the vehicle or on a test stand, fired up, and broken in.

No need to completely disassemble a donor transmission to a bare case just to cut the bellhousing end off. You only need to remove both axle side covers and remove the differential. Don't worry if it is a 6V transmission and you are starting a 12V engine. The 6V can be modified to work.

Building a Test Stand

Wouldn't it be cool to be able to start and test your engine before you install it in your car? It's not as hard as you might think to build your own VW test run stand. A few items need to be rounded up but none of it is difficult or expensive.

Once a transaxle is procured, it will need to be slightly modified. If it is complete, you will need to remove the axle side covers and differential. First drain the gearbox of all its fluids. You need a 17-mm hex wrench to do that. It's not necessary to drain it first but it keeps the mess contained.

Next remove the axle side cover nuts on both sides with a 13-mm socket. There are eight nuts on each side. From the starter side of the gearbox, tap on the differential or stub axle and the opposite side cover and differential will come out. Tap the other side cover off from the inside. Don't worry about the rest of the transmission.

With a Sawzall, cut the bellhousing off through the center section, leaving the four stud holes to mount some plates to. Cut a couple pieces of heavy-duty angle iron to span across the two farthest holes. Drill holes in the angle iron for those points.

Remove those studs and replace them with bolts.

With the angle iron bolted to the housing, weld the engine yoke that came with the engine stand to the angle iron. Do a decent job, as the entire weight of your fresh engine is going to be hanging from those welds. Everything can be blown apart, cleaned up, and painted at this point.

Jumper leads and a remote starter button are all that are necessary to make an engine run at this point, but to do a professional-looking job isn't that much work. A switch panel can be purchased from eBay for $10. It has a couple switches and a starter button. An electrical box is needed to mount the panel to. An indicator light for the oil pressure switch is added to the adapter plate for the switch panel.

You don't need to destroy a perfectly good transmission case to start an engine on the ground. This brand-new Rhino transmission case could be bolted up to an engine and a battery and starter added. Jumper leads from the battery to the coil and a remote starter button get the engine cranking and firing. Add fuel and the engine should start.

The left side of this ruler is where you want to cut the transmission case for an engine test run stand. A standard Sawzall with an aggressive blade works best. Remove the studs next to your cut line. Those 8x1.25-mm holes are the ones we use to bolt the angle iron to the case.

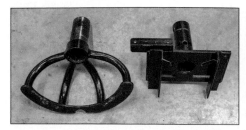

The engine stand yoke (left) is being compared to the yoke that originally came with the heavy-duty engine stand. A couple pieces of angle iron were bolted to the transmission section, centered on the yoke, and welded in place. The smaller angle iron was added to mount the electrical box.

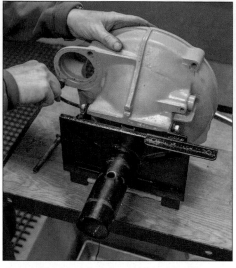

A couple pieces of angle iron were welded to the yoke that came with our Harbor Freight engine stand. They are bolted to the transmission section in four places using the 8x1.25-mm holes that originally mounted the transmission side covers.

Here is the complete engine starter adapter we made to start our engine. It consists of the V-8 engine yoke that came with our engine stand, the bellhousing section from a donor transmission, a 12V starter, a generic electrical box, a momentary switch, a toggle switch, a red indicator light, and some basic wiring.

Here is the basic layout of the run stand wiring. The lug on the starter is where all the 12V power will originate. An in-line fuse from the lug to the main power switch gets power to the box. Then we power the coil switch, starter switch, and the swap meet tachometer.

This is another example of an engine starting unit that is all self-contained. It's made to crank engines over on the ground, either to start them or at least check the compression. It has its own lawnmower fuel tank, electric fuel pump, and regulator. Clearance light serves as an oil light. Battery cables are attached, so all you need is some gas and a battery and you're in business!

Main power comes into the box via an in-line fuse from the main lug on the starter. The main power switch powers up the ignition switch, the auxiliary switch, the oil pressure light, and the tachometer. The tachometer isn't necessary, but it was a $10 swap meet purchase and it looks cool. Four wires leave the box, ignition wire powers up the coil on the (+) side, and the tach needs a signal from the coil (–) side. The ground side of the oil light goes to the oil pressure switch and the wire from the starter button goes to the starter solenoid.

Standard emergency cables will be used to connect a car battery to the positive lug on the starter and to a ground on an engine mounting stud. The beauty of using jumper cables is it won't make a difference whether the battery is in a car or not.

Installing an Engine on a Test Stand

We are going to start and test run our engine right on the engine stand we built the engine on. We will swap out the special VW-only engine yoke for the special test run adapter we made. First the engine must be removed from the build stand by simply sliding the yoke out of the stand. Again, this is a two-man job; be extra careful because it's heavy. Temporarily set the engine on the floor so you can remove the build yoke and install the test run adapter.

The adapter will cover the entire flywheel area, so don't leave pressure plate bolts partially installed or anything like that. Mount the starter adapter to the engine with all the hardware just as if you were installing the engine in a vehicle. The wiring is easy because there are only three wires. We labeled our wires with a Dymo label maker and plastic labels to make them easy to identify.

Make sure the wires aren't hanging down where they could touch the exhaust and burn. Hook jumper cables up to the starter lug (+) and engine mounting stud for the ground

(–). Hook up the other end to a car battery, either on the floor or in a car. Last, run a fuel line from the inlet on the fuel pump to a fresh can of gas. Have a fire extinguisher handy. If you are ready to start the engine now, skip ahead to the starting procedures later in this chapter.

The wire that goes from the oil pressure sender to the oil light is basically the negative side of the circuit. Power goes to one side of the oil light and is grounded out at the sender to complete the circuit. That's why the light comes on when you turn the key on. Once the engine starts and builds oil pressure, the sender loses its ground and the light goes out.

Only three wires are necessary to hook up our run stand to the engine. The oil pressure wire goes to the oil pressure switch. Coil positive (+) is switched, so 12V to the positive or 15 terminal on the coil. Coil (–) is the tach output and goes to the negative or 1 terminal on the coil.

With the engine removed from the stand, the engine build yoke is removed from the back of the engine and the engine starting adapter is installed. This gets bolted up using the same exact hardware that is used to bolt the engine in the vehicle. A 17-mm wrench is all you need.

Hook up the coil, making sure the positive (+) wire is on the 15 terminal. Hooking this wire (+) to the negative side of the coil is a surefire way to destroy the electronic ignition module. The only reason we have a negative side hookup is because we have a tachometer. The tach needs a signal from the negative (–) side of the coil to operate.

Jumper cables are hooked up to the hot side of the starter and one of the engine mounting bolts. Jumper cables are more convenient than making giant permanent leads with battery clamps. They don't limit you on what kind of battery you have to use or how far away the battery is. It could be in a vehicle for all it mattered.

Installing a 12V Starter in a 6V Transmission

You have two choices if you want to install a 12V starter into a transmission that originally housed a 6V starter. Both 6V and 12V starters are dimensionally the same except for one detail: the pilot end of the starter shaft.

Manual transmissions in almost all air-cooled Volkswagens have non-self-supporting starters, meaning the driven end of the starter shaft is supported by a bushing in the transmission housing. The 6V starters have a larger pilot shaft size, so a 12V starter will fit into the 6V transmission housing but won't be supported properly. The starter drive and flywheel will be ruined if you can even get the starter to work at all.

An easy fix to install a 12V starter into a 6V transmission is an adapter pilot bushing with a 6V OD and a 12V ID (EMPI part number 4027). Just remove the old bushing and replace it with this adapter bushing. The other option is to install a self-supporting 12V starter most commonly known as an "Auto-Stick" starter (mainly because it was installed in 1968–1972 Beetles with the "Auto-Stick" semi-automatic transmission). These starters are more expensive than the standard 12V starters, but they are a bolt-in operation for a 6V to 12V starter swap. Bosch SR17X is the part number for a self-supporting 12V starter while Bosch SR15N is the part number for a standard 12V starter.

1. Starter Comparison

Instead of going through the work to convert your 6V transmission to use a standard 12V starter (left) you can install a self-supporting starter (right). The self-supporting starter is more commonly referred to as an Auto-Stick starter (Bosch model SR17X).

2. Bushing Comparison

On the left is a standard 6V starter bushing and on the right is a standard 12V starter bushing. In the center is an adapter starter bushing that lets you install a 12V starter in a transmission originally set up for 6V starter (EMPI part number 4027).

3. 6V Bushing Removal

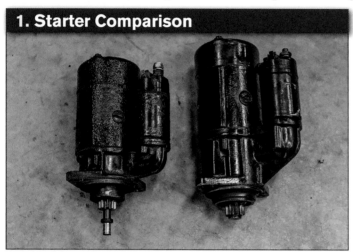

To remove the 6V starter bushing, tap the bushing, install a bolt, and pull the bolt and bushing out with the slide hammer. A 14X2.0-mm works, but if you have a 9/16-inch-12 tap and bolt that will work as well.

4. Tap Bushing

First tap the bushing with 14x2.0-mm tap. This can be done from either direction: starter side or engine side. We will show this can be done with the engine installed.

5. Use a Slide Hammer

This is a fun little tool every mechanic should create. One end is a very basic 10-inch curved jaw Vise-Grip. The adjusting screw is replaced with a 2-foot-long piece of 7/16-inch-14 threaded rod along with two nuts to lock together from the hardware store. The weight is from an old body shop slide hammer.

6. Thread the Bolt

Thread the bolt into the bushing and attach the Vise-Grip slide hammer. It shouldn't take much effort to get the bushing out; a couple of hits tops.

7. Tap in the New Bushing

Tap in the new adapter bushing with the female end of a 3/8-inch extension and a hammer. Apply a little grease to the inner diameter once it's in place.

Installing a 12V Engine in a 6V Transmission

Installing a 12V engine into a transmission originally designed for a 6V engine is a very common procedure. It goes hand in hand with installing a 12V starter and converting an entire 6V car over to 12V. Volkswagen increased the clutch diameter from 180 mm to 200 mm in 1967 when it also upped the displacement from 1,300 cc to 1,500 cc. That was also the year VW changed the electrical system over to 12V from 6V. The increase in displacement and horsepower necessitated an increase in clutch diameter.

The clutch got bigger so the flywheel got bigger. But not enough to make it impossible to fit it in. Minor clearancing is all that's required to make it fit. Once the engine is installed, all the clearancing involved is nicely hidden away inside the bellhousing.

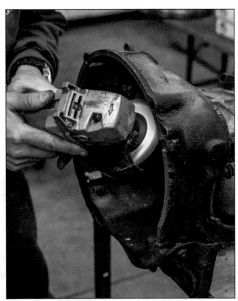

There are five areas that need to be clearanced in order to fit a 12V flywheel into a 6V transmission case. An angle grinder is all you need to remove material from the four bosses where the engine-mounting studs and bolts come through and just under the starter opening. See the shiny spots in the photo.

If you plan on clearancing 6V transmissions for 12V flywheels regularly, this little tool will come in handy. It's just the flywheel end of a junk crankshaft chopped off and turned down. The other side is drilled and tapped for a 1/2-inch stud. Attach this to a junk flywheel for test fitting.

Test fit a junk flywheel into the bellhousing opening. The flywheel is mounted to our drill arbor so it can be centered on the input shaft by the pilot bearing. It needs to go in deep enough so the back of the flywheel is slightly past flush.

Here you can see just how much material needs to be removed. The starter teeth just barely hit the casting around the four mounting holes.

The five areas have been clearanced and the test flywheel fits nicely. All the work was done with an angle grinder but a die grinder with an aggressive bit would work as well. It's definitely too much to do by hand or with a tiny Dremel-type tool.

Once you are satisfied with the fit, a quick spin with the 1/2-inch Milwaukee hole shooter will verify if enough material has been removed. The teeth on the flywheel might rub a little more material off and fine-tune the job. Just be careful and keep the RPM low.

Installing the Clutch and Pressure Plate

Clutch technology has advanced considerably since the days of the early VW engineers. Gone are three-finger-type pressure plates with a multitude of die springs to apply the clamping force. Modern pressure plates use one large diagram-type spring. The action is much smoother and the clamping force much more even across the clutch disc. They are made with fewer parts and have a lower likelihood of failure than the early multi-spring units.

Clutch material has also advanced. Gone are the soft, asbestos-type clutch discs that wore quickly, especially with an inexperienced manual transmission driver. Today's clutch discs can easily last two or three times as long as the OEM ones from the 1960s. They have metallic materials introduced into the organic friction material that is impervious to heat and heat is what usually does in a clutch.

Clutch discs and pressure plates should always be replaced as a set. In 1971, Volkswagen changed the

Sachs is a well-known OEM manufacturer of clutch kits. Here we are going to use its 200-mm rigid center clutch disc (311 141 031BK) and heavy-duty 200-mm pressure plate (311 141 025EBR). We will use a transmission input shaft (top) as a pilot tool, but EMPI sells them for $5 (3202-K).

These high-performance pressure plates show the difference between an early style (left, thru 1970) that uses a non-guided throw-out bearing and the later style (right, 1971 and on) that uses a guided throw-out bearing.

Aftermarket pressure plates from Kennedy Engineering are universal to fit both early and late-style throw-out bearings. The center collar is held in by a spiral clip that can be removed or reinstalled from the disc side of the plate.

design of its throw-out bearing, which in turn changed the design of the pressure plate. Instead of the throw-out bearing free-floating by the pivot points on the early cars, the new ones were guided by a sleeve around the input shaft. Gone was the ring on the pressure plate fingers for the throw-out bearing to mate up to. Now the throw-out bearing pushed directly on the pressure plate fingers.

1. Grease Inner Needle Bearings

Apply a small amount of grease to the inner needle bearings of the gland nut. Ours is brand new and doesn't have any lubrication whatsoever. It's important that the input shaft not bind in the gland nut, otherwise shifting through the gears may become difficult, much like the clutch not releasing completely.

2. Clean Clutch Disc

Both the flywheel and the pressure plate are brand new and are shipped with some sort of rust preventative on them. It is important to remove this with some brake cleaner so the clutch disc doesn't become contaminated and not have a chance to break in properly.

3. Insert Clutch Pilot Tool

Insert the clutch pilot tool into the clutch disc. Slide the pilot tool into the gland nut pilot bearing. Don't try to eyeball the clutch disc while installing the pressure plate without using a pilot tool. You will never get it close enough by eye.

4. Tighten Pressure Plate Bolts

Snug up each of the six bolts on the pressure plate a little at a time to apply even pressure on the disc. Occasionally make sure the pilot tool can slip in and out freely so you know the clutch disc is perfectly centered on the flywheel.

5. Torque Flywheel Bolts

Torque the flywheel bolts to 18 ft-lbs using a crisscross pattern. Double-check the centering with the pilot tool. If the clutch isn't perfectly centered on the flywheel, you will struggle to get the engine installed on the transmission, as the two need to be perfectly aligned to slide together easily.

Vehicle Installation

We will go over the installation process for the most common vehicle: the VW Beetle. It won't matter what year for the most part. The procedure will be the same.

Now is the time to tend to any issues that need attention in the engine compartment. Inspect the two rubber seals. One seal (111 813 741G) runs along the bottom of the firewall, and usually it's in good shape. The other U-shaped seal (111 813 705A or D) is usually ripped from a rough engine extraction. Replace either or both of them now, before the engine goes in.

Transmission mounts should be inspected now as well. The rear ones will rip and act like a chattering clutch disc. Stick a large pry bar between the transmission mount cradle and the transmission. It will be obvious if they are torn in two. Both sides are the same (part number 111 301 263) and fit most models 1952–1972. The engine needs to be out in order to replace these, so now is the time. The transmission nose mount should be replaced at the same time as the rear mounts. Front nose transmission mounts are part numbers 311 301 265A for 1962–1965 and 311 301 265B for 1966–1972.

Items Needed for Engine Installation

- 2 large floor jacks
- 2 jack stands
- 4x4 piece of wood roughly 2 feet long

Be sure the two upper mounting bolts are in the transmission. The right side is long and has a D-shaped head. It goes through the starter. The other is a shorter 10-mm bolt that goes next to the clutch arm. Loosen the clutch cable wing nut a few turns. Keep all the wiring clear while you are jacking up the engine. Make sure the battery is disconnected.

1. Jack up the vehicle with one floor jack (jack #1) under the rear transmission cradle. It needs to be just high enough to get the other floor jack (jack #2) under the car with the 4x4 straddling the frame horns near the front transmission mount and the ends of the floor pans.
2. Balance the engine on floor jack #1. Jack the vehicle up high enough with jack #2 so the engine can be wheeled under the rear apron.
3. Once the engine is under the vehicle and roughly centered below the engine compartment, lower the vehicle with jack #2 around the engine. Put jack stands under the torsion housing as high as they can safely go. Set the vehicle down on the jack stands but leave jack #2 under the vehicle as a safety precaution.
4. Jack the engine up with jack #1. The engine will need to be tilted slightly to get the generator pulley to clear the rubber seal. Once clear of that, it can be tilted back the other way to aim the input shaft into the pressure plate.
5. At this point, feed the throttle cable into the throttle cable tube at the back side of the fan shroud as far as you can. This will save you a lot of monkeying around later. Also push the reverse light wire through the grommet on the right side if so equipped.
6. Aim the two lower engine mounting studs into the holes in the transmission. Jack the engine up or down to make the flywheel and the mounting surface of the transmission parallel.
7. Push the engine onto the transmission. It'll take some wiggling and shoving. To help the splines of the clutch line up with the input shaft, slightly rotate the engine with a wrench. The engine and transmission need to be perfectly aligned for the two to come together.
8. Start the two lower mounting nuts (10 mm) on the studs coming through the transmission housing.
9. While still under the car, start the upper left side mounting bolt and tighten it up with a 17-mm socket and extension.
10. Tighten the two lower mounting nuts with a 17-mm wrench.
11. Attach the rubber fuel line to the hard line coming out of the left side frame horn.
12. Push the heater box bellows onto the heater boxes on both sides.
13. Make sure the throttle cable made it through the tube and the end is showing. It's safe to remove jack #1 now.
14. From the top, reach behind the right side of the fan shroud with a 10-mm nut and start it on the fourth engine mounting stud. Tighten it up with a 17-mm ratcheting wrench.
15. Attach the throttle cable to

the carburetor and tighten up the barrel nut with an 8-mm wrench. Have someone go inside the car and floor the gas pedal. Check for full throttle on the carb end.

16. The wiring on the left side is the black wire to the (+) or 15 terminal on the coil and the blue with green wire to the oil pressure switch.

17. The wiring on the right side is the larger red wire that goes to D+ on the generator and the green wire to DF on the generator. (On 6V vehicles, these wires go to the regulator; on one-wire alternator vehicles

the large red one goes from the battery to the alternator post. The blue wire from the indicator light in the speedometer goes to the one terminal on the alternator.)

18. Reverse lights (if equipped) get power from the (+) or 15 terminal on the coil and use an in-line fuse. That black wire connects to the wiring that comes through a grommet behind the fan shroud on the left side.

19. Heater box cables that exit both frame horns under the car attach to the arms on the heater boxes with barrel nuts.

20. The apron tin attaches with four 6-mm screws and seals the underside of the vehicle from the engine compartment.

21. The heater tubes can be attached from the fan shroud to the muffler or directly to the heater boxes, depending on the exhaust system.

22. Adjust the clutch pedal via the wing nut at the transmission. You are looking for 1/4- to 1/2-inch free play at the top of the pedal.

23. The vehicle can come off the jack stands now.

24. Reconnect the vehicle's battery.

Start the Engine

We are starting our engine on a test stand, but the procedure is exactly the same if you installed it in a vehicle. Unless you have a remote starter button, getting the engine started in a vehicle is going to be a two-man job. Make sure your battery is fully up to the task. Put a battery charger on it if necessary.

We need to crank the engine over to prime the oiling system. Sometimes it takes a while. Keep the distributor clamp just tight enough to turn it by hand.

1. Remove all four spark plugs. This will allow the engine to crank faster, building oil pressure sooner.

2. Remove the wire from the 15 terminal on the coil if in a vehicle or turn the ignition switch off on the test stand.

3. Turn the key to the on position in a vehicle or switch the main power on the stand. The oil light should light up.

With the battery hooked up, flip on the main power switch. The oil light should light up. Hit the ignition switch. You should hear the fuel shut off solenoid on the carburetor click. Turn everything back off if these things are in order.

4. Crank the engine over until the oil light goes out. This might take a while, so go in 20- to 30-second bursts. Crank until the light goes out and then 10 seconds more.

5. Once the oil light goes out, the oil system is primed. Reinstall the spark plugs and spark plug wires. Connect a timing light.

Connect the timing light to the battery and number-1 spark plug wire. Run a fuel line from the inlet side of the fuel pump to a can of fresh gas. Our test run exhaust system connects to a piece of flex pipe and into a cherry bomb glass pack muffler and out underneath the door of the shop.

Remove all of the spark plugs. With the main power on but the ignition off, crank the engine over until the oil light goes out. This may take some time, so only crank in 30-second intervals as to not overheat the starter. Reinstall all the spark plugs and spark plug wires.

6. Connect the black wire to the 15 terminal on the coil or switch on the ignition on the stand.

7. All this cranking should've primed the fuel system. If not, carefully pour some fuel down the vent tube of the carburetor.

Fill the float bowl of the carburetor with a little gas. You can pour it right down the vent tube. The cap from a can of brake cleaner works great for this. Keep the fire extinguisher handy. A common problem is a backfire through the carburetor caused by the timing being off.

8. Pump the gas a couple times and try to start the engine. If it doesn't start, turn the distributor a few degrees and try again.

9. If it still doesn't start, check your spark plug wires and make sure they are going to the right spark plugs. The number-1 wire lines up with the notch in the distributor under the cap. Then it's 1-2-3-4 in a counterclockwise rotation around the cap.

10. Once it's started, turn the distributor to get the highest idle. Set the timing light to 28 degrees. Using the zero dimple on the outside of the pulley, set the timing so it lines up with the split in the case at maximum advance. All the advance should be in by 2,500 rpm. Don't worry about what the timing is at idle.

Time to light this thing up! Have the distributor clamp just tight enough that you can move the distributor by hand, the main power on, and the ignition on. Pump the throttle a couple times and crank it over. If it doesn't start right away, twist the distributor a couple degrees and try again. Once it starts, twist the distributor until it idles as fast as possible.

Set your timing light to 28 degrees. Using the zero dimple on the outside of the pulley, rotate the distributor until the dimple is lined up with the split in the case while revving the engine over 2,500 rpm. Don't worry about what the timing is at idle. We are looking for a maximum of 28 degrees no matter how high it revs.

11. If you rebuilt your engine with a new cam and lifters, now is the time to break them in. Most cam manufacturers recommend running the engine 15 to 20 minutes while varying the RPM between 2,000 and 3,000. Make sure you start off with enough fuel to do this. Keep one eye glued to the oil light and the other looking for oil leaks from below.

12. After the break-in period, change the oil. Dump the oil into a light-colored pan if you have one. The oil should be red in color from the assembly lube. A few tiny particles floating about is alright, but if it looks like metal flake paint then something went wrong.

13. Fill the engine with fresh oil (and filter if so equipped) and readjust the valves to a tight 0.006 inch. (See chapter 7 if you need to refresh your memory on that procedure.)

14. Start the engine for the second time and warm it up. Adjust

the idle using the bypass screw (larger one) on the left side of the carburetor. Set the volume control screw (smaller one) next to the bypass screw by turning out for maximum RPM then in until the RPM drops by 20 to 30 rpm. Readjust the idle to 850 to 900 rpm.

15. Double-check the maximum timing is 28 degrees total advance at over 2,500 rpm and lock down the distributor clamp.

We knew we had oil pressure but just how much was uncertain. We unscrewed the oil pressure switch for the light and screwed in a portable oil pressure gauge. We had 40 psi at idle, which is plenty. The factory spec is 42 psi at 2,500 rpm with the oil temp at 158°F.

With the engine warmed up, we can make the final carburetor adjustments. The volume control screw is the smaller one on the left side. Turn it out until it idles the fastest then slowly turn it in until the engine speed drops 20 to 30 rpm. The larger screw on the left side is the idle bypass. Set the idle to 850 to 900 rpm.

You can make a simple oil pressure tester like this out of spare parts laying around. It's an in-dash mechanical oil pressure gauge, some 1/8-inch pipe fittings, and a rubber flex brake line from the front end of a Beetle.

Once the carb is adjusted, double-check the timing again. You should have 28 degrees at maximum advance. Lock down the distributor clamp now. With the electronic ignition that's the last time you should ever have to set the timing.

Last, we topped off the carburetor with a new chrome aftermarket air cleaner. A 1/2-inch breather hose was cut and runs from the bottom of the air cleaner to the fitting next to the oil fill cap.

Parts Cost Breakdown

Long-Block			
Item	**Manufacturer**	**Part Number**	**Cost**
85.5-mm piston and barrel kit	Moresa	311-198-069FD	$149.95
Pushrod tubes (8)	EMPI	311-109-335	$16.00
Gasket kit	Elring	111-198-007AFG	$18.00
Rear seal	Elring	113-105-245SF	$3.50
Oil pressure switch	EMPI	98-9190	$4.95
Main bearings	Silver Line	111-198-473BR	$47.99
Rod bearings	Silver Line	113-105-707	$14.95
Cam bearings	Silver Line	111-198-541T	$14.99
Reground lifters (8)	DPR Machine	–	$35.00
Connecting rods	EMPI	311-105-401B	$74.95
Oil pump	EMPI	00-9207	$59.95
Flywheel	EMPI	311-105-273A	$42.95
Gland nut	EMPI	00-4029	$16.00
Oil strainer	OEM	111-115-175B	$4.50
Long-Block Total:			$503.68

Parts Cost Breakdown *Continued*

Accessories

Item	Manufacturer	Part Number	Cost
34PICT carburetor	EMPI	98-1289	$99.95
Electronic ignition	EMPI	00-9432	$34.95
Spark plugs (4)	Bosch	W8AC	$8.00
Distributor cap	Bosch	03 010	$13.50
Distributor rotor	Bosch	04 110	$11.50
Spark plug wires	EMPI	98-9925	$29.95
Doghouse fan shroud	EMPI	00-8672	$59.95
Intake manifold boots	EMPI	113-129-729BRS	$6.00
Fan belt	Continental	9.5x905 mm	$7.95
55-amp alternator kit	EMPI	00-9445	$99.95
Air cleaner	EMPI	17-2978	$13.95
Fuel pump	EMPI	113-127-025G	$19.95
Fuel pump rod	EMPI	113-127-307A	$3.95
Fuel line 5-mm 5 foot	OEM	N203551	$9.95
Fuel filter	EMPI	131-261-275	$1.95
Accessory Total:			$421.45

Machine Shop Bill

Item	Manufacturer	Part Number	Cost
Line bore case +0.020			$65.00
Install 10-mm case savers			$85.00
10-mm case savers (2)	EMPI	4011	$15.90
Flycut heads (2)			$80.00
Clean heads and cut valve seats, install valves (2 heads)			$80.00
New intake valves (4)	EMPI	311-109-601	$24.00
New exhaust valves (4)	EMPI	311-109-612A	$24.00
Regrind crankshaft -0.010			$75.00
Machine Shop Total:			$448.90
Long-block parts			$503.68
Accessories			$421.45
Machine shop			$448.90
Grand Total:			$1,374.03
(All prices are as of August 16, 2018)			

Not included in this breakdown are common items you may or may not have already, such as oil, oil filter, assembly lube, sealers, spray paint, etc. Also not factored in are specialty tools that you may or may not have or are going to borrow or rent. This breakdown will show you how your build may compare to an engine that is already rebuilt.

Most rebuilt VW engines require a core charge as well. The core charge can range anywhere from $250 to $450. This must be factored into the grand total. Rebuildable cores are getting harder and harder to come by.

Most turnkey rebuilt VW engines are priced complete minus exhaust and clutch. The price of exhaust components vary greatly depending on what route you are taking. A stock muffler with chrome exhaust tips is approximately $110. New heater boxes are approximately $150 per side. Exhaust systems that eliminate the heat are obviously much cheaper. This break down is also minus exhaust and clutch. ∎

Cam Timing Alignment Marks

With the crank set into the 3-4 side of the case, the cam timing is set by lining up the single mark on the cam gear between the two marks on the crank gear. Once aligned, mesh the gears together while rotating the cam into the cam bearings. See page 75 for more details.

Firing Order

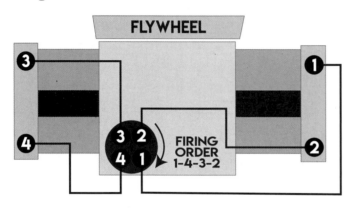

Torque Specifications

Location	Torque	Wrench Size
Connecting rod nuts	25 ft/lbs	14 mm
Case perimeter nuts	14 ft/lbs	13 mm
Main bearing nuts	25 ft/lbs	17 mm
Cylinder head nuts (8 mm)	18 ft/lbs	15 mm
Cylinder head nuts (10 mm)	23 ft/lbs	15 mm
Rocker shaft to head	18 ft/lbs	13 mm
Oil pump nuts	14 ft/lbs	13 mm
Oil drain plug	25 ft/lbs	13/16 inch
Oil strainer sump nuts	5 ft/lbs	10 mm
Flywheel to crankshaft	253 ft/lbs	36 mm
Crankshaft pulley	36 ft/lbs	30 mm
Clutch to flywheel	18 ft/lbs	13 mm
Spark plugs	25 ft/lbs	13/16 inch
Generator pulley	43 ft/lbs	21 mm
Special nut for fan	43 ft/lbs	36 mm
Oil cooler to case (Early)	5 ft/lbs	10 mm
Oil cooler to case (Late)	14 ft/lbs	13 mm

Head Torque Pattern

Initial

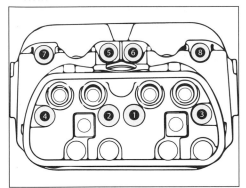

Torque the head to 7 ft-lbs using the sequence shown. This procedure is to properly crush the pushrod tubes.

Final

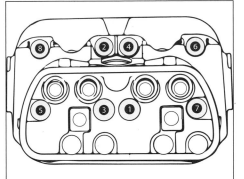

The final head torque sequence is different from the initial torque sequence. It basically moves from the center out. Torque the 10-mm head studs to 23 ft-lbs (8-mm head studs get torqued to 18 ft-lbs).

Compression Ring Gap

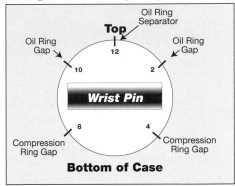

Index the piston rings as shown in this diagram.

Tune-Up Specifications

Ignition timing with factory distributors, vacuum hose(s) off, 800 to 900 rpm idle	
1,200 cc	10 degrees before TDC
1,300 cc	7.5 degrees before TDC
1,500 cc	0 degrees
1,600 cc (single-port)	0 degrees
1,600 cc (dual-port)	5 degrees after TDC at 3,500 rpm
Ignition Timing with mechanical advance distributors	
All years/engine sizes	26-28 degrees total advance at 2,500 to 3,500 rpm
Spark Plug Gap	0.026 to 0.028 inch
Recommended Spark Plugs	Bosch W8AC or NGK B5HS
Point Gap	0.016 to 0.020 inch
Ignition Dwell Angle	44 to 50 degrees (wear limit 42 to 58 degrees)
Valve Lash	0.006-inch intake and exhaust (engine cold)
Oil capacity	2.65 quarts (3.0 with a filter pump)
Oil viscosity	10W40 climates under 32°F
	20W50 climates over 32°F
Oil filter for filter pump	EMPI 9208
	Fram PH2870A
	WIX 51393
	ACDelco PH56
	NAPA 1342

SOURCE GUIDE

AA Performance Products
228 S. Fifth Ave.
City of Industry, CA 91746
626-333-5555
aapistons.com

CB Performance
1715 N. Farmersville Blvd.
Farmersville, CA 93223
800-274-8337
cbperformance.com

California Import Parts LTD
800-313-3811 (US and Canada)
800-628-596 (Australia)
cip1.com
cip1.ca in Canada

California Pacific/JBugs Inc.
1338 Rocky Point Dr.
Oceanside, CA 92056
800-231-1784
jbugs.com

Car Craft Inc.
1006 E. La Cadena
Riverside, CA 92507
951-781-0452
carcraftstore.com

The Dub Shop
17239 Tye St. S.E., Ste. E
Monroe, WA 98272
206-414-8456
thedubshop.com

Dune Buggy Warehouse
2610 Bobmeyer Rd.
Hamilton, OH 45015
513-868-9543
dunebuggywarehouse.com

EMPI Inc.
301 E. Orangethorpe Ave.
Anaheim, CA 92801
800-666-3674
empius.com

Fat Performance - Rimco Machine
1558 N. Case St.
Orange, CA 92867
714-637-2889
fatperformance-rimco.com

Fisher Buggies
5006 E. Acline Dr.
Tampa, FL 33619
800-741-2841
fisherbuggies.com

Mid-America Motorworks
2900 North Third St.
Effingham, IL 62401
866-350-4535
mamotorworks/vw.com

SCAT Enterprises, Inc.
1400 Kingsdale Ave.
Redondo Beach, CA 90278
310-370-5501
scatenterprises.com

The Samba Online Community
thesamba.com

West Coast Metric
24002 Frampton Ave.
Harbor City, CA 90710
800-247-3202
westcoastmetric.com

Wolfsberg West
2850 Palisades Dr.
Corona, CA 92880
888-965-3937
wolfsbergwest.com

Magazines
Airmighty Magazine
airmighty.com

Hot VWs magazine
1520 Brookhollow Dr., #39
Santa Ana, CA 92705
714-979-3998
hotvws.com

Volks America Magazine
volksamerica.com

Volks World Magazine
Hill View, Thetford Road, Northwold,
Thetford, Norfolk, IP26 5LQ UK
Tel: 01366 728488
volksworld.com